Botany FOR DUMMIES®

by René Fester Kratz, PhD
Biology professor, Everett Community College,
Everett, Washington

Wiley Publishing, Inc.

Botany For Dummies®

Published by
Wiley Publishing, Inc.
111 River St.
Hoboken, NJ 07030-5774
www.wiley.com

Copyright © 2011 by Wiley Publishing, Inc., Indianapolis, Indiana

Published by Wiley Publishing, Inc., Indianapolis, Indiana

Published simultaneously in Canada

No part of this publication may be reproduced, stored in a retrieval system or transmitted in any form or by any means, electronic, mechanical, photocopying, recording, scanning or otherwise, except as permitted under Sections 107 or 108 of the 1976 United States Copyright Act, without either the prior written permission of the Publisher, or authorization through payment of the appropriate per-copy fee to the Copyright Clearance Center, 222 Rosewood Drive, Danvers, MA 01923, (978) 750-8400, fax (978) 646-8600. Requests to the Publisher for permission should be addressed to the Permissions Department, John Wiley & Sons, Inc., 111 River Street, Hoboken, NJ 07030, (201) 748-6011, fax (201) 748-6008, or online at http://www.wiley.com/go/permissions.

Trademarks: Wiley, the Wiley Publishing logo, For Dummies, the Dummies Man logo, A Reference for the Rest of Us!, The Dummies Way, Dummies Daily, The Fun and Easy Way, Dummies.com, Making Everything Easier, and related trade dress are trademarks or registered trademarks of John Wiley & Sons, Inc. and/or its affiliates in the United States and other countries, and may not be used without written permission. All other trademarks are the property of their respective owners. Wiley Publishing, Inc., is not associated with any product or vendor mentioned in this book.

LIMIT OF LIABILITY/DISCLAIMER OF WARRANTY: THE PUBLISHER AND THE AUTHOR MAKE NO REPRESENTATIONS OR WARRANTIES WITH RESPECT TO THE ACCURACY OR COMPLETENESS OF THE CONTENTS OF THIS WORK AND SPECIFICALLY DISCLAIM ALL WARRANTIES, INCLUDING WITHOUT LIMITATION WARRANTIES OF FITNESS FOR A PARTICULAR PURPOSE. NO WARRANTY MAY BE CREATED OR EXTENDED BY SALES OR PROMOTIONAL MATERIALS. THE ADVICE AND STRATEGIES CONTAINED HEREIN MAY NOT BE SUITABLE FOR EVERY SITUATION. THIS WORK IS SOLD WITH THE UNDERSTANDING THAT THE PUBLISHER IS NOT ENGAGED IN RENDERING LEGAL, ACCOUNTING, OR OTHER PROFESSIONAL SERVICES. IF PROFESSIONAL ASSISTANCE IS REQUIRED, THE SERVICES OF A COMPETENT PROFESSIONAL PERSON SHOULD BE SOUGHT. NEITHER THE PUBLISHER NOR THE AUTHOR SHALL BE LIABLE FOR DAMAGES ARISING HEREFROM. THE FACT THAT AN ORGANIZATION OR WEBSITE IS REFERRED TO IN THIS WORK AS A CITATION AND/OR A POTENTIAL SOURCE OF FURTHER INFORMATION DOES NOT MEAN THAT THE AUTHOR OR THE PUBLISHER ENDORSES THE INFORMATION THE ORGANIZATION OR WEBSITE MAY PROVIDE OR RECOMMENDATIONS IT MAY MAKE. FURTHER, READERS SHOULD BE AWARE THAT INTERNET WEBSITES LISTED IN THIS WORK MAY HAVE CHANGED OR DISAPPEARED BETWEEN WHEN THIS WORK WAS WRITTEN AND WHEN IT IS READ.

For general information on our other products and services, please contact our Customer Care Department within the U.S. at 877-762-2974, outside the U.S. at 317-572-3993, or fax 317-572-4002.

For technical support, please visit www.wiley.com/techsupport.

Wiley also publishes its books in a variety of electronic formats. Some content that appears in print may not be available in electronic books.

Library of Congress Control Number: 2011930301

ISBN 978-1-118-00672-6 (cloth); ISBN 978-1-118-11082-9 (ebk); ISBN 978-1-118-11083-6 (ebk); ISBN 978-1-118-11084-3 (ebk)

Manufactured in the United States of America

10 9 8 7 6 5 4 3 2 1

About the Author

René Fester Kratz, Ph.D., grew up near the ocean in Rhode Island. From a young age, she wanted to be a teacher (because she loved her teachers at school) and a biologist (because her dad was one). She graduated from Warwick Veterans Memorial High School and then went on to major in biology at Boston University. As she studied biology at BU, René was most interested by the stories about plants, fungi, and microbes — organisms that are traditionally studied under the umbrella of botany. After René graduated with a B.A. in Biology from BU, she worked for two years in the Department of Botany at the University of Massachusetts in Amherst and then went on to get an M.S. and Ph.D. in botany from the University of Washington. At UW, René studied reproductive onset in *Acetabularia acetabulum,* a marine green alga that grows as single cells big enough to pick up with your fingers. When they enter reproduction, the cells of *A. acetabulum* form a flat disk or cup-shaped structure at the top, earning the alga the nickname of "The Mermaid's Wineglass."

René currently teaches biology and general science classes at Everett Community College in Everett, Washington. René spends most of her time introducing students to the wonders of cells and microbes as she teaches cellular biology and microbiology, but she's never forgotten her love of plants. René also has a strong interest in science education and science literacy for everyone. As a member of the North Cascades and Olympic Science Partnership, she helped create inquiry-based science courses for future teachers that are based on research on human learning. René loves teaching these courses because they make science accessible for all kinds of people. In the summer, René enjoys working with K–12 teachers on the improvement of science education in the public schools. René also enjoys writing about science and is the author of *Molecular & Cell Biology For Dummies, Biology For Dummies,* 2nd Edition, and *EZ Microbiology.*

René loves living in the Pacific Northwest because she is near the ocean and her daffodils start blooming in February (when her family back East is still shoveling snow). She doesn't mind the rain and thinks the San Juan Islands are one of the most beautiful places on earth. Her husband, two sons, and two very bad dogs help her remember what is truly important, and her "sisters" help keep her sane. René loves to scrapbook, quilt, stitch, garden, and read.

Dedication

To the people who got me excited about plants and fungi: Drs. Gillian Cooper-Driver, Richard B. Primack, Walt Halperin, Fayla Schwartz, and Elizabeth VanVolkenburgh. And to my sister, Alyson, who helped me with my first plant collection when I was an undergrad.

Author's Acknowledgments

Thanks to Matt Wagner, of Fresh Books, Inc., for helping me find the opportunity to write this book. And thanks to all the great people at Wiley who made it happen: my editor, Kelly Ewing, who was always helpful and upbeat; the acquisitions editor, Stacy Kennedy, who helped get me started on the project; Alicia South, who coordinated the art; and Fayla Schwartz, Ph.D., and Linda Sigismondi, Ph.D., my technical reviewers. Thanks also to Sheree Montgomery, the project coordinator, and Kathryn Born, Karl Brandt, and Mark Pinto, who worked on the art.

A very special thanks to my colleague, Dr. Fayla Schwartz, for reviewing this book and also for teaching me about the plants of the Pacific Northwest when we were both graduate students at the University of Washington.

On the home front, thanks to my husband, Dan, for all of his love and support. To Hueston and Dashiel for once again being patient when Mommy was glued to her computer.

Thanks to all of my students at Everett Community College for your enthusiasm and hard work. You have all inspired me to keep doing what I do. Thanks also to my Dean, Al Friedman, for letting me teach a reduced load for one quarter, so I'd have more time to write.

Publisher's Acknowledgments

We're proud of this book; please send us your comments at `http://dummies.custhelp.com`. For other comments, please contact our Customer Care Department within the U.S. at 877-762-2974, outside the U.S. at 317-572-3993, or fax 317-572-4002.

Some of the people who helped bring this book to market include the following:

Acquisitions, Editorial, and Media Development

Project Editor: Kelly Ewing

Acquisitions Editor: Stacy Kennedy

Assistant Editor: David Lutton

General Reviewers: Fayla Schwartz, PhD, and Linda Sigismondi, PhD

Senior Editorial Manager: Jennifer Ehrlich

Editorial Supervisor and Reprint Editor: Carmen Krikorian

Editorial Assistant: Rachelle S. Amick

Art Coordinator: Alicia B. South

Cover Photos: (Front cover) © iStockphoto.com/Andres Peiro Palmer; (Back cover) © iStockphoto.com/Oliver Sun Kim, © iStockphoto.com/Natalia Kuzmina

Cartoons: Rich Tennant (www.the5thwave.com)

Composition Services

Project Coordinator: Sheree Montgomery

Layout and Graphics: Kim Tabor

Proofreaders: Jessica Kramer, BIM Indexing & Proofreading Services

Indexer: Cheryl Duksta

Special Art: Karl Brandt, Mark Pinto

Special Help
Illustrations by Kathryn Born, M.A.

Publishing and Editorial for Consumer Dummies

    Kathleen Nebenhaus, Vice President and Executive Publisher

    Kristin Ferguson-Wagstaffe, Product Development Director

    Ensley Eikenburg, Associate Publisher, Travel

    Kelly Regan, Editorial Director, Travel

Publishing for Technology Dummies

    Andy Cummings, Vice President and Publisher

Composition Services

    Debbie Stailey, Director of Composition Services

Contents at a Glance

Introduction 1

Part I: Plant Basics 5
Chapter 1: Exploring Botany 7
Chapter 2: Peering at Plant Cells 13
Chapter 3: Identifying Plant Tissues 43
Chapter 4: Vegetative Structures: Stems, Roots, and Leaves 57
Chapter 5: Reproductive Structures: Spores, Seeds, Cones, Flowers, and Fruits ... 81

Part II: The Living Plant: Plant Physiology 99
Chapter 6: Metabolism: How Living Things Get Energy and Matter 101
Chapter 7: Photosynthesis: Making Sugar from Scratch 109
Chapter 8: Cellular Respiration: Making Your Cake and Eating It, Too 131
Chapter 9: Moving Materials Through Plants 147
Chapter 10: Regulating Plant Growth and Development 163

Part III: Making More Plants: Plant Reproduction and Genetics 179
Chapter 11: Greening the Earth: Plant Reproduction 181
Chapter 12: Passing Plant Characteristics to the Next Generation 199

Part IV: The Wide, Wonderful World of Plants: Plant Biodiversity 215
Chapter 13: Changing with the Times: Evolution and Adaptation 217
Chapter 14: The Tree of Life: Showing the Relationships Between Living Things 231
Chapter 15: Examining the Forest Floor: Bryophytes and Seedless Vascular Plants 243
Chapter 16: Their Seeds Are Naked: Gymnosperms 265
Chapter 17: Say It with Flowers: Angiosperms 275

Part V: Plants and People 289
Chapter 18: Making Connections with Plant Ecology 291
Chapter 19: Altering Plants by Using Biotechnology 311
Chapter 20: Thriving on Plants in Everyday Life 323

Part VI: The Part of Tens 341
Chapter 21: Ten Weirdest Plants 343
Chapter 22: Ten Tips for Improving Your Grade in Botany 349

Index 355

Table of Contents

Introduction 1
About This Book 1
Conventions Used in This Book 1
What You're Not to Read 2
Foolish Assumptions 2
How This Book Is Organized 2
    Part I: Plant Basics 2
    Part II: The Living Plant: Plant Physiology 3
    Part III: Making More Plants: Plant Reproduction and Genetics 3
    Part IV: The Wide, Wonderful World
        of Plants: Plant Biodiversity 3
    Part V: Plants and People 3
    Part VI: The Part of Tens 4
Icons Used in This Book 4
Where to Go from Here 4

Part 1: Plant Basics 5

Chapter 1: Exploring Botany 7
Taking a Close Look at Plant Structure 7
    Organizing plants into roots, stems, and leaves 8
    Finding ways to procreate 8
Figuring Out Plant Functions 9
    Making and using food 9
    Transporting materials 10
    Responding to hormones 10
Considering Plant Reproduction and Genetics 11
Exploring the Wide World of Plants 11
Making Connections Between Plants and People 12

Chapter 2: Peering at Plant Cells 13
Making Molecules from Matter 13
    Examining elements, atoms, and isotopes 14
    Getting atoms together to form molecules 17
Making Acids and Bases 18
Building Cells from Four Types of Molecules 19
    Carbohydrates 20
    Proteins 22
    Nucleic acids 23
    Lipids 26

Entering the World of Cells 29
    Customs: Plasma membrane 30
    The library: Storing information in DNA 31
    Factories: Ribosomes 32
Exploring Plant Cells 33
    Post office: The endomembrane system 35
    Scaffolding and railroad tracks: The cytoskeleton 36
    Solar-powered batteries: Chloroplasts 37
    Powerplants: Mitochondria 39
    Rebar and concrete: Cell walls and extracellular matrices 40

Chapter 3: Identifying Plant Tissues 43

Combining Cells to Form Tissues 43
    Growing with meristems 44
    Building plant bodies with ground tissue 47
    Transporting materials with vascular tissue 52

Chapter 4: Vegetative Structures: Stems, Roots, and Leaves 57

Getting Organized into Plant Organs 57
Getting Taller with Stems 58
    Primary growth 59
    Secondary growth 63
    Specialized stems 66
Digging Deep with Roots 68
    Regions within roots 69
    Specialized roots 72
    Forming partnerships with soil fungi 74
Reaching Out with Leaves 75
    Leaf structure 75
    Leaf types 77
    Leaf arrangements 78
    Specialized leaves 79

Chapter 5: Reproductive Structures: Spores, Seeds, Cones, Flowers, and Fruits 81

Reproducing with Spores 81
Protecting the Offspring with Seeds 82
    Seed structure 83
Organizing Reproduction in Cones 87
Finding a Mate with Flowers 87
    Flower structure 87
    Moving pollen 90
    Flower arrangement 90
Packaging the Seeds in Fruits 92
    Fruit types 93
Dispersing Fruits and Seeds 97

Part II: The Living Plant: Plant Physiology 99

Chapter 6: Metabolism: How Living Things Get Energy and Matter 101
- The Big Picture: Overview of Metabolism 101
- Moving through Metabolic Pathways 103
- Speeding Things Up with Enzymes 104
- Transferring Energy with ATP 105
- Shuttling Electrons with Electron Carriers 107

Chapter 7: Photosynthesis: Making Sugar from Scratch 109
- Rethinking the Role of Soil 109
  - van Helmont's experiment 110
  - The real role of soil 111
- Discovering Photosynthesis Fundamentals 112
  - Going solar 113
  - Catching some rays with pigments 114
  - Connections between the light and light independent reactions 116
- Harnessing the Sun: The Light Reactions 118
  - Transferring light energy to chemical energy with electron transport chains 118
  - Noncyclic and cyclic photophosphoryation 122
- Storing Matter and Energy in Sugar: The Light Independent Reactions 125
  - The steps of the light independent reactions 125
  - Getting carbon dioxide to rubisco 126

Chapter 8: Cellular Respiration: Making Your Cake and Eating It, Too 131
- Digging into Cellular Respiration Fundamentals 131
- Breaking Down Glucose in Glycolysis 134
  - Glycolysis is universal 135
  - Making ATP by substrate-level phosphorylation 135
  - The steps of glycolysis 136
- Going Farther with the Krebs Cycle 139
  - More is better 139
  - The steps of the Kreb's cycle 141
- Making Useful Energy: Chemiosmosis and Oxidative Phosphorylation 143
  - Transferring electrons 144
  - Transferring energy 145

Chapter 9: Moving Materials Through Plants ... 147
Shipping and Receiving Materials throughout the Plant ... 147
Moving Across Membranes ... 148
  Diffusion ... 149
  Active transport ... 150
  Osmosis ... 151
  Under pressure ... 152
  Losing it ... 154
Going with the Flow: Water Transport ... 154
  Clinging for support: The cohesion of water ... 155
  The pull from above: Transpiration ... 156
  Fixing a break in the lines: Cavitation ... 158
  Reaching out with roots ... 159
Sticky Business: Sugar Transport ... 160
  Sources and sinks ... 160
  Pressure-flow hypothesis ... 161

Chapter 10: Regulating Plant Growth and Development ... 163
Getting an Overview of Plant Growth and Development ... 163
  Receiving signals ... 164
  Responding to signals ... 164
Sending Signals with Plant Hormones ... 165
  Auxins ... 166
  Cytokinins ... 169
  Gibberellins ... 170
  Abscisic acid ... 170
  Ethylene ... 171
  Brassinosteroids ... 171
Which Way Do I Go?: Plant Movements ... 172
  Growth movements (tropisms) ... 172
  Turgor movements ... 173
What Time Is It?: Sensing the Seasons ... 174
  Flowering ... 175
  Circadian rhythm ... 176
  Seed germination ... 177

Part III: Making More Plants: Plant Reproduction and Genetics ... 179

Chapter 11: Greening the Earth: Plant Reproduction ... 181
Reproducing: More Than One Way to Do It ... 181
  Asexual reproduction ... 182
  Sexual reproduction ... 184
  Comparing reproductive styles ... 184

Copying Cells by Mitosis 185
    Interphase 186
    Overview of mitosis 187
    Cytokinesis 189
Reproducing Sexually with Meiosis 189
    Counting chromosomes 190
    Following the plan 191
    Doubling your stuff and then reducing it twice 191
    The events of meiosis I 192
    The events of meiosis II 194
Considering Alternation of Generations 194

Chapter 12: Passing Plant Characteristics to the Next Generation 199

Tracking the Inheritance of a Single Gene 199
    Investigating plants with Gregor Mendel 200
    Figuring out the rules of inheritance 202
    Speaking like a geneticist 204
    Making predictions 205
Tracking the Inheritance of Two Independent Genes 207
    Adding plant height to the mix 207
    Solving the Punnett square for a dihybrid cross 209
    Remembering meiosis 211
Mixing it Up with Incomplete Dominance 212

Part IV: The Wide, Wonderful World of Plants: Plant Biodiversity 215

Chapter 13: Changing with the Times: Evolution and Adaptation ... 217

Figuring Out the Fundamentals of Evolution 217
    Mutation 218
    Natural selection 218
Identifying Important Factors in Plant Evolution 222
    Hybridization 222
    Polyploidy 223
    Reproductive isolation 223
Admiring Plant Adaptations 223
    Desert plants 224
    Tropical rainforest plants 226
    Carnivorous plants 228
    Aquatic plants 229

Chapter 14: The Tree of Life: Showing the Relationships Between Living Things 231

Examining the Branches of the Tree of Life 231
  Digging into the three domains 232
  Exploring the past through phylogenetic trees 234
Organizing Life 236
  Developing a system 236
  Defining plants 239
Naming the Rose (And Other Living Things) 240

Chapter 15: Examining the Forest Floor: Bryophytes and Seedless Vascular Plants 243

Moving onto the Land 244
Bryophytes: Nonvascular Plants 246
  Liverworts: Phylum Hepatophyta 246
  Hornworts: Phylum Anthocerophyta 249
  Mosses: Phylum Bryophyta 250
Seedless Vascular Plants 253
  Club mosses: Phylum Lycophyta 255
  Ferns and (some of) their allies: Phylum Pterophyta 260

Chapter 16: Their Seeds Are Naked: Gymnosperms 265

Protecting the Embryo with Seeds 265
Cycads: Cycadophyta 268
Ginkgoes: Ginkgophyta 269
Conifers: Coniferophyta (Pinophyta) 270
  Pines 271
Gnetophytes: Gnetophyta 274

Chapter 17: Say It with Flowers: Angiosperms 275

Delving into Flowering Plants: Anthophyta 275
  Searching for the origin of the angiosperms 275
  Discovering the unique characteristics of angiosperms 278
  Looking at the life cycle of angiosperms 279
Exploring Angiosperm Diversity 281
  Basal angiosperms 282
  Magnoliids 282
  True dicots 282
  Monocots 285
Thinking about Pollination Ecology 286

Part V: Plants and People 289

Chapter 18: Making Connections with Plant Ecology 291
Exploring Ecosystems 291
    Figuring out your job description 292
    Going with the flow of energy 293
    Going round and round with matter cycles 298
Interacting with Other Organisms 302
    Living together 302
    Working things out 303
Peeking at Powerful Plant Communities 304
    Exploring biomes 305
    Changing communities through succession 307

Chapter 19: Altering Plants by Using Biotechnology 311
Oh, No! GMOs! 311
Engineering Plants 312
    Culturing plant tissues 313
    Transforming plants with Agrobacterium 314
Trying to Build a Better World 317
    Saving lives with golden rice 318
    Stopping crop pests with "Bt" 318
    Resisting weed killers 319
    Making drugs and commercial enzymes with pharma crops 320
Let's Talk, Talk, Talk about It: Pros
and Cons of Genetic Engineering 320

Chapter 20: Thriving on Plants in Everyday Life 323
Feeding a Hungry World 323
    Exploring the origins of agriculture 323
    Keeping up with human population growth 325
Making Use of Plant Products 327
    Building homes 327
    Making paper 328
    Wearing cotton 330
    Fueling the future 330
Discovering the Power of Plant Chemicals 333
    Finding a cure: Medicinal plants 333
    Dangerous weapons: Poisonous plants 335
    Having visions: Hallucinogenic plants 337

Part VI: The Part of Tens 341

Chapter 21: Ten Weirdest Plants 343

Plants That Eat Insects: Cobra Lily 343
Plants That Stink: The Corpse Flower 344
Plants That Move: Galloping Moss 344
Plants That Mimic Rocks: Stone Plants 344
Plants That Come Back From the Dead: Resurrection Plant 345
Plants That Just Look Strange: Welwitschia 345
Plants That Aren't Green: Indian Pipe 346
Plants That Have Sex with Insects: Australian Tongue Orchid 346
Plants That Really Know How to Grow: Queen Victoria Water Lily 347
Plants That Climb: Banyan Tree 348

Chapter 22: Ten Tips for Improving Your Grade in Botany 349

Listen Actively to Lecture 349
Use Your Lab Time Effectively 350
Plan Your Study Time 351
Be Active, Not Passive 352
Make Up Tricks to Jog Your Memory 352
Prepare for Different Kinds of Questions 353
Remember the Supporting Material 353
Test Yourself Often 354
Use Your First Test as a Diagnostic Tool 354
Get Help Sooner Rather Than Later 354

Index 355

Introduction

*E*verybody's talking about green these days. Green technology, green energy, green lifestyles. It's no accident that the word green has come to symbolize a healthy world and sustainable habits. People use the word green because it ties into visions of a green and growing world, lush with forests and fields of plants. This book is about the organisms that form the foundation of that green vision — the plants that surround us, support us, and make your world beautiful.

About This Book

Botany For Dummies is an introduction to the world of plants and their importance to the rest of life on earth. My goal is to present the concepts of plant biology in a clear and straightforward way, while I help you relate the science to your everyday life. I include lots of pictures of the processes, structures, and life cycles that you'd typically encounter in an introductory course in botany.

Botany is the study of plants, which covers a wide range of subjects, including their structure, function, patterns of inheritance, diversity, and importance to humans. I hope you'll be as surprised and intrigued as I was when I first began to study botany and realized that the seemingly simple world of plants was actually pulsing with life, mystery, and beauty.

Conventions Used in This Book

In order to explain things as clearly as possible, I keep scientific jargon to a minimum and present information in straightforward style. I break dense information into main concepts and divide complicated processes into series of steps.

To help you find your way through the subjects in this book, I use the following style conventions:

- *Italic* is used for emphasis and to highlight new words or terms that are defined in the text.
- **Boldface** is used to indicate key words in bulleted lists or the action parts of numbered steps.

- Sidebars are shaded gray boxes that contain text that's interesting to know but not critical to your understanding of the topic.
- Web addresses are written in monofont so that they're easy to recognize.

What You're Not to Read

Sidebars are shaded gray boxes that include stories or information related to the main topic, but not necessary to your understanding. You can skip the sidebars if you want. Any information marked with a Technical Stuff icon can also be skipped without hurting your understanding of the main concepts.

Foolish Assumptions

As I wrote this book, I tried to imagine who you might be and what you might need in order to understand botany:

- A college biology major studying botany as part of your year-long freshman series.
- A college student taking an introductory botany class as a way to fulfill the science requirement for your degree in a nonscience field.
- Someone who just wants to know a little bit more about plants – you may be a gardener or a hiker who enjoys the beauty of plants and wants to know a little bit more about how they work.

How This Book Is Organized

I organized the major subtopics in botany — plant structure (morphology), function (physiology), patterns of inheritance (genetics and development), diversity (taxonomy and phylogeny), and importance to humans —into parts.

Part 1: Plant Basics

Plants have many similarities to other types of life on earth. Just like all organisms, the structure and function of plants begins at the level of the cell. Plant cells combine to form tissues, which then form familiar plant organs like leaves, flowers, and fruits. In this part, I cover the plant cell and then show you how these cells are organized to form the basic plant structures.

Part II: The Living Plant: Plant Physiology

Plants have the same basic needs as other organisms: They need a source of building material (matter) and energy to grow and function. Of course, plants are special because they have the ability to collect energy from the sun and use it to build their own food, which they can then use as a source of matter and energy for growth. Because plants are pretty much stuck in one place, they also have unique ways of dealing with the issues of getting and storing water. This part begins with an overview of the fundamentals of plant metabolism and then presents the mechanisms that plants use to transport materials and respond to environmental signals.

Part III: Making More Plants: Plant Reproduction and Genetics

Plants grow and reproduce, both sexually and asexually. In this part, I show you how plant cells divide, either to make exact copies of themselves for growth or to make eggs and sperm for sexual reproduction. I'll also walk you through basic genetics so you can follow how plants pass characteristics from one generation to the next.

Part IV: The Wide, Wonderful World of Plants: Plant Biodiversity

You're probably familiar with some of the plants that grow in your world, especially those in your yard, neighborhood, or on your dining table. But I wonder if you've ever considered the mosses, grasses, evergreens and flowers in detail, and thought about what makes each of them unique. In this part, I introduce the major plant groups, presenting the structures and life cycles that make each group unique.

Part V: Plants and People

Plants are interwoven throughout the fabric of human lives, from their importance to the health of our environment to the products that people make from them. In this section, I show you the importance of plants to the natural world, then explore the many ways that people make use of plants, including for food, clothing, fuel and medicine. I also explain how people are using DNA technology to modify plants, changing our agricultural practices and developing medicines for the poor.

Part VI: The Part of Tens

This part of *For Dummies* books contains short chapters containing lists of ten or so items. In this book, I give you ten strange but true plant stories and ten ways to improve your grade in botany.

Icons Used in This Book

The familiar *For Dummies* icons are used here to help guide you and give you new insights as you read the material.

This bull's-eye symbol lets you know what you need to do to get to the heart of the matter at hand. These icons mark information that helps you remember the facts being discussed or suggest a way to help you commit it to memory.

This information gives you extra information that isn't necessary to understand the topic. If you want to take your biology learning to a higher level, incorporate these paragraphs into your reading. If you want just the basics and do not want to be confused by the details, skip them.

This little icon serves to jog your memory. The information spotlighted here is information you should permanently store in your biology file. If you want a quick review of biology, scan through the book reading the remember icons. No need for a chunky yellow highlighter.

Where to Go from Here

Like all *For Dummies* books, each chapter in *Botany For Dummies* is self-contained, so you can pick up whenever you need it and jump into the topic you are working on. You can start anywhere in the book that you want. If you are reading this book for general interest, you might enjoy starting with the last section on Plants and People. If you're taking a college class in botany, you'll probably want to start at the beginning with the basics on plant cells, tissues, and organs.

I hope you enjoy your journey into the world of plants and find them as amazing and beautiful as I do!

Part I
Plant Basics

The 5th Wave   By Rich Tennant

"That should do it."

In this part . . .

This part introduces you to some of your neighbors on planet earth that you might not have noticed much before. Plants seem pretty different from people, and although many people appreciate the beauty of plants, not everyone takes the time to really get to know them.

In this part, you take a closer look and discover that plants have many similarities to you. For example, like all living things, plants are made of cells, tissues, and organs. Plants even have sex! In fact, many of the foods you eat are the result of plant sexual reproduction. You can find out all about these similarities as you read this part, in which I present the fundamentals of plant structure that are essential to understanding plant function and evolution.

Chapter 1

Exploring Botany

In This Chapter
▶ Building plants one cell at a time
▶ Finding out about how plants work
▶ Connecting plants and people

*B*otany is the study of plants, including plant structure, function, reproduction, diversity, inheritance, and more. Plants may seem like they're part of the background of your life, when really they're at the center. The food you eat, the clothes you wear, the materials that make up your home — all these things depend upon plants. Plants remove carbon dioxide from the atmosphere, helping to keep your planet from getting too warm for life as you know it. They provide homes for insects and other animals, filter impurities out of ground water, and help protect shorelines from erosion.

And beyond all these useful things plants do, they're just cool! Plants have many unique strategies that help them survive in all different kinds of environments. They trap and trick insects, grow in the ground or up in the rainforest canopy, and manage to survive everywhere from the glacial arctic to the hot, dry deserts. They seem so different from people, and yet when you really look at how plants grow and function, you'll be surprised at how similar they are to you. This chapter offers an overview of the science of botany, giving you a peek into the mysteries of plants.

Taking a Close Look at Plant Structure

You might not think so, but plants are a lot like you. Their bodies are made of cells (see Chapter 2) that are organized into tissues (see Chapter 3) that form the familiar plant organs of roots, stems, and leaves (see Chapter 4). Plant cells use the same basic chemistry as your cells, storing information in DNA, using carbohydrates for energy, and putting proteins to work. And your cells and plant cells are both *eukaryotic cells,* meaning they have a similar structure that includes a nucleus and cellular organelles.

Organizing plants into roots, stems, and leaves

Plants reach out to the sun with their leaves, absorbing light energy so that they can make sugar through the process of photosynthesis. Most leaves are flat because that's the best shape for spreading out and catching lots of sun. But plants can also make leaves in different shapes for different purposes, such as tendrils for hanging on and climbing, spines for protecting the plant against grazing animals, or thick, fleshy leaves for storing water.

Plant stems support the leaves, holding them in different arrangements so that they don't shade each other and can absorb the most light. New plant growth occurs at stem tips as cells divide to make stems grow longer and to build new leaves, branches, and sometimes flowers. Some plant stems, such as those in cacti, are green so that they can do photosynthesis. Other types of stems, such as the runners of a strawberry plant, grow along the ground, sending up new plants at intervals along the horizontal stem.

Plant roots are in charge of getting water for the leaves and stems by absorbing water from the soil. Along with the water, plant roots absorb minerals that provide them with the nitrogen, phosphorous, and other elements they need to function. Some plants, such as dandelion, have long tap roots that reach down deep into the soil, while others, such as grass, have many small fibrous roots. Plants like corn may make extra roots, called *prop roots,* that start from the stem above ground and then grow down to sink into the earth.

Finding ways to procreate

Plants have many ways of reproducing themselves. When plants reproduce sexually, they make special reproductive cells called *spores* (see Chapter 5). Many familiar plants make a structure that's even better at starting the next generation — the *seed*. Seeds protect the plant embryos they carry and nourish them with stored food.

Some plants that do sexual reproduction get fancy and produce showy flowers (see Chapter 5) designed to attract animals to help spread their pollen around. Other flowering plants just dangle their flowers in the wind and let the wind do the work.

Flowers contain the male and female parts of the plant that will participate in sexual reproduction.

Pollen comes from the male part of flowers, carrying and protecting the plant sperm. The female parts of flowers house the ovules that contain the eggs. *Pollination* occurs when pollen arrives at the female part of the flower. The pollen releases the sperm so that they can fuse with the egg, causing *fertilization,* and starting the next plant generation. After fertilization in flowering plants, the ovaries within the flowers develop into fruits (see Chapter 5). Some fruits are sweet and fleshy, inviting animals to come eat the fruit and then disperse the seeds. Other fruits are dry and designed to either float on the breeze, hitch a ride on some animal fur, or even explode to release their seeds. Whatever the method, the goal is the same — to find a nice, new home for the embryos inside the seed to grow.

Figuring Out Plant Functions

In addition to being made of cells and having similar chemistry, plants use many of the same strategies that you do to solve life's challenges. Both you and plants need a source of building material, called *matter,* to build the cells of your body, and you both need a source of *energy* so that you can build things and move around (see Chapter 6). And just like you, plants need to transport food and fluids around their bodies. Finally, you and plants both grow and develop, responding to changes in your environment.

Making and using food

The go-to source of matter and energy for all living things is food. Of course, one big difference between you and a plant is that you have to get your food by eating another organism, while plants can make their own.

Plants make their own food through the process of *photosynthesis* (see Chapter 7). Although the process of photosynthesis is pretty complex, you can get the main idea if you think of it like a recipe. The ingredients are carbon dioxide from the atmosphere and water taken up from the soil. You then follow these directions:

1. **Use light energy from the sun to combine carbon dioxide and water, rearranging the atoms to form sugar and oxygen.**

2. **Serve sugar to all parts of the plant that need matter and energy and throw the oxygen gas away.**

3. **If you have leftovers, you can combine the sugars into starch to store some for later.**

When plants want to use some of the sugar they've made to provide themselves with matter and energy, their cells do the same thing that your cells do with food, they break it down in a process called *cellular respiration* (see Chapter 7). Cellular respiration is a series of chemical reactions that basically unpack food molecules, making the matter and energy available to cells. When cells use cellular respiration to extract all the energy they can from food molecules, they release the waste matter as carbon dioxide and water.

Transporting materials

All the cells of a plant need food to provide them with matter and energy. Plants usually make sugars in their leaves, so they have to ship those sugars from the leaves to the rest of the plant. Likewise, plants take in water through their roots, but they need to get water to the entire plant, especially to the leaves where it's needed for photosynthesis. So, just like you have veins and arteries to transport blood around your body, plants have vascular tissue that specializes in the transport of sugar and water (see Chapter 9).

Plants transport dissolved sugars using a special type of tissue called *phloem*, and they transport water and dissolved minerals using a tissue called *xylem*.

Phloem transports sugar from the leaves where it's made through photosynthesis, to all parts of the plant that need it for growth or that will store it as starch for later. Xylem transports water from the roots up through the plant to supply all the cells with the water they need.

Responding to hormones

Yet another similarity between you and plants is that they use *hormones* to direct their growth and development (see Chapter 10).

Although plants never go through puberty (lucky plants!), they do undergo major developmental changes such as when a seed switches from being dormant to beginning to grow, or when a flowering plant decides it's the right time of year to start putting on a floral display. Plant hormones also direct responses like helping plants shoots grow toward the light and causing plant roots to grow downward toward the pull of gravity.

Considering Plant Reproduction and Genetics

Plants grow like, well, like weeds. That's because weeds are plants. (Okay, now I'm just being silly.) But seriously, plants grow when groups of cells at their tips, called *apical meristems,* divide in two to produce new cells. The process of cell division that adds new growth is called *mitosis* (see Chapter 11). Plants do mitosis pretty much the same way your cells do. Woody plants also do mitosis to grow wider, adding girth to tree trunks.

Plants also reproduce sexually, combining sperm and egg to make the next generation of plants. The sperm and egg, which are produced by a type of cell division called *meiosis* (see Chapter 11), carry copies of the DNA from the parent plant, passing their traits onto their offspring.

By following the inheritance of traits from one generation to the next through the science of *genetics* (see Chapter 12), scientists can figure out how plant genes interact with each other to determine the traits of a plant.

Exploring the Wide World of Plants

Planet earth is filled with a glorious diversity of plants. Plants can be as tall as the mighty redwood tree, or as small as the tip of a pin. They can grow so rapidly that they go from seed to seed in a month, or they can live for over a thousand years. Since plants moved onto the land over 400 million years ago, they've evolved to live in every type of environment (see Chapter 13): Today, plants grow in the deserts, in the rainforests, in the ocean, and up on the mountain. Some plants make flowers, while others make cones. Plants may trap insects as a source of minerals or lure them in to help with pollination. With all these different strategies and environments, you can probably imagine that some pretty amazing plants are out there, from delicate mosses (see Chapter 15), to sturdy pine trees (see Chapter 16) and colorful flowers (see Chapter 17).

Botanists study all the different kinds of plants to understand how each one gets what it needs to survive and reproduce. They also compare the structures and DNA code of plants to figure out the relationships between plant groups and reconstruct how plants evolved (see Chapter 14).

Making Connections Between Plants and People

The lives of people are completely interwoven with the lives of plants:

- **Plants support the ecosystems of which people are a part.** Without plants to supply food to the web of life, what would you eat? (For more on this topic, see Chapter 18.)

- **People can modify plants to make them more nutritious or so they produce medicines.** Genetically modified foods are very controversial, but they have benefits as well as risks (see Chapter 19).

- **People grow plants for food.** The origins of human agriculture stretch back 10,000 years. And the switch from hunting and gathering to farming changed the entire structure of human societies (see Chapter 20).

- **People use plants to make clothing.** Cotton and flax plants are used to make cotton and linen clothing. Some of the dyes people use to give their clothing color also come from plants. (see Chapter 20).

- **People get medicines from plants.** Digitalin for heart disease, aspirin to reduce fever, and artemisinin for malaria are just a few examples of the powerful drugs people have extracted from plants(see Chapter 20).

- **People use plants for building materials.** People use wood for houses, furniture, and tools, or straw as material for roofs or bricks.

- **People reduce their stress and improve their fitness by taking a walk and admiring the plants.** Seriously, reducing stress is important. Stress has major impacts on people's health. And for many people, nature has a soothing effect.

- **Plants help keep water clean.** You probably hear people talking about wetlands, how they're important, and how they're disappearing at a rapid rate, thanks to development. Wetlands are communities with certain types of plants and soils. As the rain falls across areas where humans live, it picks up lawn fertilizers, motor oil from cars, poop from pets, and more. If this runoff flows through a wetland before it enters our streams and lakes, the plants and bacteria in the wetlands will remove lots of the dangerous substances on the way. Having wetlands to slow the flow of water also helps prevent flooding.

However you look at it, from the similarities of plants to other organisms, their beauty, or their usefulness to humans, you can certainly find lots of good reasons to know a little bit more about plants.

Chapter 2
Peering at Plant Cells

In This Chapter
- Building molecules from atoms
- Discovering the molecules that make up the cell
- Exploring plant cells

All living things, including plants, are made of cells. Plant structure and function depend on these cells, so it's a good idea to make sure that you have a firm grasp of cellular basics before you dive into the details of botany presented in the rest of this book. In this chapter, I give you an overview of the fundamentals of cells and the molecules they're made of.

Making Molecules from Matter

Matter is anything that takes up space and can be weighed — in other words, it's the stuff that makes up the living and nonliving things on planet earth.

Just like you, the air you're breathing, and the book you're holding, plants are made of matter. You're probably familiar with some of the molecules that make up matter on earth — the proteins, fats, and carbohydrates in your body, the cellulose in the paper of the book, the carbon dioxide and oxygen in the air you're breathing. Plants are made of the same kinds of molecules as you are, and they exchange carbon dioxide and oxygen with the atmosphere, too. The following sections describe these molecules, and the atoms that make them up.

Examining elements, atoms, and isotopes

Different kinds of materials, called *elements,* make up matter. You're probably familiar with many common elements like copper, iron, chlorine, and calcium. *Elements* are pure substances because they're made up of all the same kind of small particles, called atoms. *Atoms* are the smallest pieces of an element that have the characteristics of the element. For example, the element copper is a shiny metal that conducts electricity and interacts with other elements in a certain way. Iron is also a metal element, but it has different properties than copper. If you could isolate a single atom from either copper or iron metal, the individual atoms, which are so tiny you can't even see them with a microscope, would still have the same properties as the larger metal they came from.

All the elements that have been identified so far are organized into a table called The Periodic Table of Elements (see Figure 2-1):

- **Each row of the table is called a *period.*** The elements in a period get heavier as you move across the table horizontally.
- **Each column is called a *family* or *group.*** Elements within the same family/group have similar properties. The size of the atoms gets larger as you move from top to bottom within each column.

1 H Hydrogen 1.01																	2 He Helium 4.00
3 Li Lithium 6.94	4 Be Beryllium 9.01											5 B Boron 10.81	6 C Carbon 12.01	7 N Nitrogen 14.01	8 O Oxygen 16.00	9 F Fluorine 19.00	10 Ne Neon 20.18
11 Na Sodium 22.99	12 Mg Magnesium 24.31											13 Al Aluminum 26.98	14 Si Silicon 28.09	15 P Phosphorus 30.97	16 S Sulfur 32.06	17 Cl Chlorine 35.45	18 Ar Argon 39.95
19 K Potassium 39.10	20 Ca Calcium 40.08	21 Sc Scandium 44.96	22 Ti Titanium 47.90	23 V Vanadium 50.94	24 Cr Chromium 52.00	25 Mn Manganese 54.94	26 Fe Iron 55.85	27 Co Cobalt 58.93	28 Ni Nickel 58.71	29 Cu Copper 63.55	30 Zn Zinc 65.38	31 Ga Gallium 69.72	32 Ge Germanium 72.59	33 As Arsenic 74.92	34 Se Selenium 78.96	35 Br Bromine 79.90	36 Kr Krypton 83.80
37 Rb Rubidium 85.47	38 Sr Strontium 87.62	39 Y Yttrium 88.91	40 Zr Zirconium 91.22	41 Nb Niobium 92.91	42 Mo Molybdenum 95.94	43 Tc Technetium (99)	44 Ru Ruthenium 101.07	45 Rh Rhodium 102.91	46 Pd Palladium 106.42	47 Ag Silver 107.87	48 Cd Cadmium 112.41	49 In Indium 114.82	50 Sn Tin 118.69	51 Sb Antimony 121.75	52 Te Tellurium 127.60	53 I Iodine 126.90	54 Xe Xenon 130.30
55 Cs Cesium 132.91	56 Ba Barium 137.34	57 La Lanthanum 138.91	72 Hf Hafnium 178.49	73 Ta Tantalum 180.95	74 W Tungsten 183.85	75 Re Rhenium 186.21	76 Os Osmium 190.2	77 Ir Iridium 192.22	78 Pt Platinum 195.09	79 Au Gold 196.97	80 Hg Mercury 200.59	81 Tl Thallium 204.37	82 Pb Lead 207.19	83 Bi Bismuth 208.98	84 Po Polonium (210)	85 At Astatine (210)	86 Rn Radon (222)
87 Fr Francium (223)	88 Ra Radium (226)	89 Ac Actinium (227)	104 Rf Rutherfordium (257)	105 Db Dubnium (260)	106 Sg Seaborgium (263)	107 Bh Bohrium (262)	108 Hs Hassium (265)	109 Mt Meitnerium (266)									

Lanthanides	58 Ce Cerium 140.12	59 Pr Praseodymium 140.91	60 Nd Neodymium 144.24	61 Pm Promethium (147)	62 Sm Samarium 150.35	63 Eu Europium 151.96	64 Gd Gadolinium 157.25	65 Tb Terbium 158.93	66 Dy Dysprosium 162.50	67 Ho Holmium 164.93	68 Er Erbium 167.26	69 Tm Thulium 168.93	70 Yb Ytterbium 173.04	71 Lu Lutetium 174.97
Actinides	90 Th Thorium (232)	91 Pa Protactinium (231)	92 U Uranium (238)	93 Np Neptunium (237)	94 Pu Plutonium (242)	95 Am Americium (243)	96 Cm Curium (247)	97 Bk Berkelium (247)	98 Cf Californium (251)	99 Es Einsteinium (254)	100 Fm Fermium (257)	101 Md Mendelevium (258)	102 No Nobelium (259)	103 Lr Lawrencium (260)

Figure 2-1: The Periodic Table of the Elements.

Elements and atoms

Each element is different from the other elements because of the structure of its atoms. Atoms are made of smaller components (see Figure 2-2A):

A. Bohr's model of an atom: carbon used as an example.

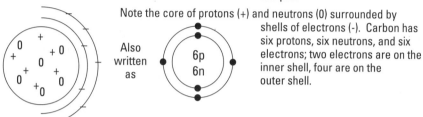

Note the core of protons (+) and neutrons (0) surrounded by shells of electrons (-). Carbon has six protons, six neutrons, and six electrons; two electrons are on the inner shell, four are on the outer shell.

Also written as

B. Sodium and chloride ions joining to form table salt. The sodium ion has a positive charge because there's one more proton than electrons, so the overall charge is positive. The chloride ion is negative because after it accepts the electron from sodium, it then has one more electron than protons (18 versus 17), so the overall charge is negative. Together, though, NaCl is neutral because the "plus 1" charge is balanced by the "minus 1" charge.

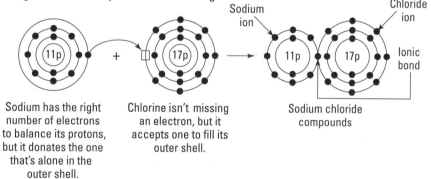

Sodium has the right number of electrons to balance its protons, but it donates the one that's alone in the outer shell.

Chlorine isn't missing an electron, but it accepts one to fill its outer shell.

Sodium chloride compounds

C. Two atoms of oxygen joining to form oxygen gas.

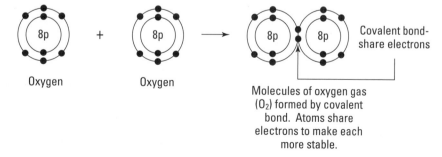

Oxygen   Oxygen

Molecules of oxygen gas (O_2) formed by covalent bond. Atoms share electrons to make each more stable.

Covalent bond - share electrons

Figure 2-2: Atoms and chemical bonds.

- ✓ **Protons and neutrons make up the core of an atom, called the *atomic nucleus*.** Both of these particles have mass and contribute to the weight and size of the atom. Protons have a positive electrical charge, while neutrons have no charge.
- ✓ **Electrons are negatively charged particles that form a cloud around the atomic nucleus.** Electrons have no mass.

Ions and isotopes

The number of protons in an atom, called the *atomic number*, determines the type of element. All iron atoms (atomic symbol: Fe), for example, have 26 protons, while all copper atoms (atomic symbol: Cu) have 29 protons.

While the number of protons in an element are always the same, the numbers of neutrons or electrons may change:

- ✓ **When an atom gains or loses electrons, it becomes an *ion*.** In an atom of an element, the number of positively charged protons is equal to the number of negatively charged electrons. The opposite electrical charges cancel out each other, and the atom has no net charge. If an atom gives up electrons to another atom, then the atom will have a net positive charge, becoming a positively charged ion. Likewise, if an atom gains electrons from another atom, then the atom will have a net negative charge, becoming a negatively charged ion. For example, when sodium and chlorine atoms interact, sodium gives an electron to chlorine (see Figure 2-2B), forming a positively charged sodium ion and a negatively charged chloride ion.
- ✓ **Atoms with the same number of protons, but different numbers of neutrons, are *isotopes* of each other.** Neutrons have mass, so isotopes of an element are heavier or lighter than each other. Atoms of the element carbon (atomic symbol: C), for example, always have 6 protons — that's what makes them carbon atoms. Most carbon atoms also have 6 neutrons, giving them a total mass, called their *mass number*, of 12. Some carbon atoms, however, have 6 protons and 8 neutrons, giving them a mass number of 14. Thus, carbon-12 and carbon-14 are isotopes of each other.

Scientists have estimated the ratio of different isotopes of each element found on Earth and used that ratio to calculate the average *atomic mass* of each element. Most carbon atoms on Earth have a mass number of 12, but a small fraction have other mass numbers, like 13 and 14. If you average the mass numbers, taking into account the proportion of the different isotopes, you get an atomic mass for carbon of 12.01. So, in any sample containing carbon atoms, you'd expect the mass of the carbon atoms to average out at 12.01.

Getting atoms together to form molecules

Atoms join together to form *molecules,* chemical structures made of two or more atoms. The way an atom interacts with other atoms depends upon the number of electrons it has. Atoms will give, take, or share electrons with other atoms in order to achieve a stable electron arrangement. The attractions that join atoms together to form molecules are called *chemical bonds.*

Four types of chemical bonds are particularly important in plant cells:

- **Ionic bonds** are the electrical attractions between oppositely charged ions (see Figure 2-2B). Ionic bonds can be very strong in dry environments like salt crystals, but they're weak in the watery world of plant cells.

- **Covalent bonds** form when atoms share electrons (see Figure 2-2C). Covalent bonds are the strongest bonds in plant cells, creating the backbones of the molecules that form the cell.

- **Polar covalent bonds** are a special category of covalent bonds, forming when atoms share electrons unequally. In water molecules, for example, the oxygen nucleus has a stronger attraction for the shared electrons than does the hydrogen nucleus. The electrons spend more time hanging around the oxygen side of the molecule, giving that side a slight negative charge. Due to the absence of the electrons, the hydrogen side of the molecule has a slight positive charge (check out Figure 2-3).

- **Hydrogen bonds** are weak electrical attractions between polar groups. Polar covalent bonds create polar molecules that have slight electrical charges. You can think of these molecules like they're electrically sticky — positive sides of one molecule are attracted to the negative sides of other molecules (like the water molecules in Figure 2-3, where $\delta+$ marks the weak positive charges near the hydrogen atoms, and $\delta-$ marks the weak negative charge near the oxygen atom.). The hydrogen bonds between water molecules give water many of its unique properties, like the fact that ice floats or that liquid water has surface tension. Hydrogen bonds between polar groups within a molecule can also help give individual molecules useful shapes.

18 Part I: Plant Basics

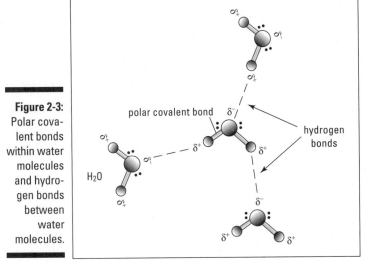

Figure 2-3: Polar covalent bonds within water molecules and hydrogen bonds between water molecules.

Making Acids and Bases

Molecules can change when they're mixed with water to make a *solution*, and can even change the properties of the solution itself.

Molecules called *acids* and *bases* release ions into a solution. When solutions become acidic or basic, they can damage cells:

- **Acids are molecules that release hydrogen ions (H^+) into solutions.** When hydrochloric acid (HCl) is added to a solution, it splits apart into hydrogen ions (H^+) and chloride ions (Cl^-), increasing the number of hydrogen ions in the solution.

- **Bases are molecules that release hydroxide ions (OH^-) into solutions or that remove hydrogen ions (H^+) from the solution.** When the base sodium hydroxide is added to a solution, it splits apart into sodium ions (Na^+) and hydroxide ions (OH^-). The hydroxide ions can combine with hydrogen ions in the solution, forming water.

The ions that acids and bases release into solution can interfere with the chemical bonds that hold molecules together, leading to cell damage. You can measure the potential for cellular damage by looking at the *pH* of a solution, which is a measure of the concentration of hydrogen ions (H^+) in the solution. Increasing hydrogen ions makes a solution more *acidic,* while decreasing hydrogen ions makes a solution more *basic*.

The *pH scale*, shown in Table 2-1, shows the pH of some common substances. Most cells, including yours and the cells of plants, have a *neutral* pH right in the middle of the scale at pH7. Acidic substances like lemon juice have a lower pH, while basic substances like oven cleaner have a higher pH. The farther the pH of a solution gets from pH 7, the greater the potential that it can disrupt the molecules that make up cells and do cellular damage.

Table 2-1   The pH of Common Substances

pH Value	Substance
0 (most acidic)	Hydrochloric acid (HCl)
1	Battery acid
2	Lemon juice, vinegar, stomach acid
3	Cola, apples
4	Beer
4.5	Tomatoes
5	Black coffee, bananas
5.5	Normal rainwater
6	Urine
6.5	Saliva, milk
7 (neutral)	Pure water, tears
7.5	Human blood
8	Seawater, eggs
9	Baking soda, antacids
10	Great Salt Lake
11	Ammonia
12	Bicarbonate of soda, soapy water
13	Oven cleaner, bleach
14 (most basic)	Sodium hydroxide (NaOH)

Building Cells from Four Types of Molecules

Molecules are the building blocks that make up cells. You can think of molecules like little chemical Legos that are arranged and rearranged to build the structure of each living and growing cell. In multicellular organisms like plants, cells join together to form the tissues that make up the structure of the organism.

REMEMBER

The cells of all living things, including plant cells, are primarily made of four types of big molecules, called *macromolecules*:

- ✔ Carbohydrates
- ✔ Proteins
- ✔ Nucleic acids
- ✔ Lipids

Carbohydrates

Carbohydrates are commonly referred to as sugars, and the foods you can think of that are naturally sweet — like fruit, for example — are probably high in carbohydrates. Cells use carbohydrates for storing energy and building materials and also to provide structure to the cell.

Several types of carbohydrates are important to cells:

- ✔ **Many sweet tasting carbohydrates are smaller, simple sugars, called *monosaccharides* by scientists.** Glucose is an example of a monosaccharide. Glucose is extremely valuable to cells because it can be used as a fast source of energy. Glucose can exist in the linear form shown in Figure 2-4a, but in the water-filled environment of a cell, the molecule loops around and binds to itself, forming a ring-shaped structure (shown in Figure 2-4b).

- ✔ **Monosaccharides may form bonds with each other to form larger structures.**

    - When glucose bonds with fructose, the sugar found in fruit, they form the *disaccharide* sucrose, otherwise known as common table sugar (see Figure 2-4b). Plants make sucrose in their green structures and then ship it all around the plant body to provide matter and energy to all their cells.

    - Short chains of monosaccharides are called *oligosaccharides* (see Figure 2-4c). Oligosaccharides send signals to plant cells, triggering growth responses and defense mechanisms.

    - Long chains of monosaccharides form *polysaccharides* (see Figure 2-4d. You may have heard polysaccharides referred to as complex carbohydrates. Like monosaccharides, polysaccharides are important molecules for storing energy and building materials and then making them available to cells. For example, the starch found in rice, pasta, breads, and potatoes is a polysaccharide that's an important source of energy for both plants and people. Plants reinforce the structure of their cells with the polysaccharide cellulose, which is one of the major components of the cell wall that surrounds and supports plant cells.

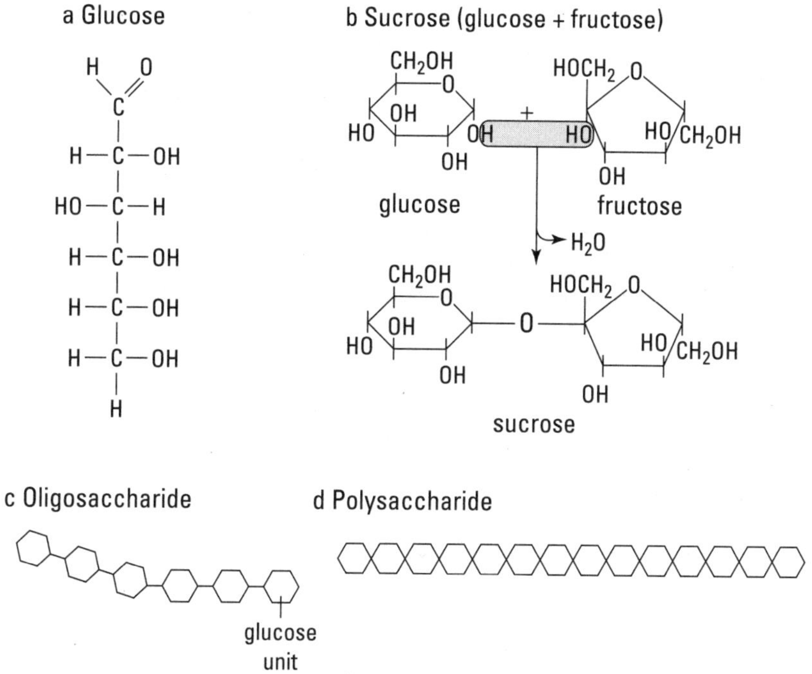

Figure 2-4: Carbohydrates.

a Glucose

b Sucrose (glucose + fructose)

c Oligosaccharide — glucose unit

d Polysaccharide

Get your fiber here!

Fruits and vegetables, as well as other plant foods like nuts and whole grains, are an excellent source of fiber. But what is fiber? And why do plants have it? *Fiber* is the common name for a plant polysaccharide called cellulose that surrounds and supports plant cells. Just like the starch you eat as a source of energy and building materials for our cells, cellulose is a long chain of glucose molecules strung together. However, there's an important difference between the way the glucose molecules are attached to each other in starch and the way they're attached together in cellulose. Well, this difference is important to people anyway, because humans can break the links in starch molecules, but not in cellulose molecules. So, when you digest starch, your body separates it into individual glucose molecules that you can easily break down for energy or rearrange to build your cells. But when cellulose, or fiber, hits your digestive system, it just passes on through as long chains. Your body can't access the glucose molecules at all. So, all that undigested fiber passes into your large intestine and helps give mass to your, er, waste, which helps keep your large intestine healthy and functioning normally. Fiber can help lower blood cholesterol, control blood sugar, and help people lose weight. So, be sure to include plenty of plants in your daily diet!

Proteins

Plant cells couldn't function without proteins. That's because proteins perform essential jobs in cells, moving materials around, supporting the cell, helping chemical reactions, controlling information flow, and sending signals.

Each protein has a unique shape that helps it do its job. To make a protein, cells link *amino acids* with covalent bonds called *peptide bonds* (shown in Figure 2-5), forming long chains of amino acids called *polypeptide chains*. The polypeptide chains fold up, either singly or in groups, to form the final shape of the functional protein.

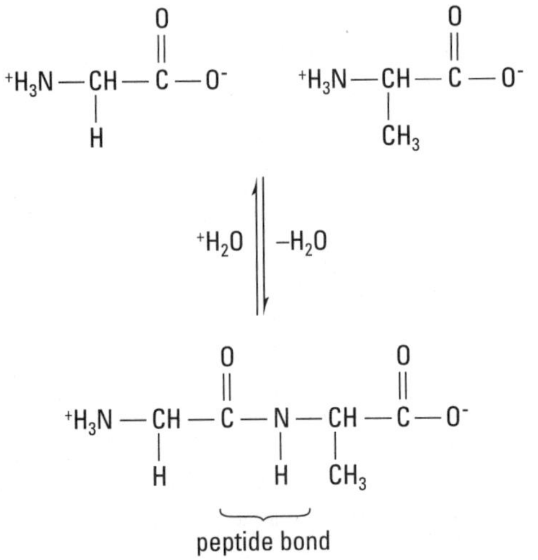

Figure 2-5: Amino acids link together to form proteins.

Proteins have so many functions in plant cells that a list could go on for two pages. Rather than overwhelm you with all those functions at once, I hit a few of the most important functions here and then introduce specific proteins as they're needed throughout the book:

- ✔ **Enzymes are proteins that speed up chemical reactions.** As they live and grow, plants are constantly building new molecules and breaking other molecules down. The speed of these chemical reactions by themselves wouldn't be fast enough to keep up with the pace of life. So plant cells, just like all cells, use enzymes to make those reactions happen exactly when plants need them.

- ✔ **Structural proteins support the cell.** Protein cables inside plant cells, called *cytoskeletal proteins,* provide supportive scaffolding from the inside. (For more details on cytoskeletal proteins, see the upcoming section "Scaffolding and railroad tracks: The cytoskeleton.") Outside the cell, proteins are woven into the plant *cell wall*, a protective layer that encases plant cells. (You can find out more about plant cell walls in the section "Rebar and concrete: Cell walls and extracellular matrices.")

- ✔ *Transport proteins* **move materials into and within plant cells.** Plants need to move molecules in and out of their cells. Transport proteins located at the boundary of the cell help create passageways for these materials. Inside the cell, molecules and structures may use cytoskeletal proteins as tracks that allow them to move around the cell.

- ✔ *Receptor* **proteins help plant cells communicate.** In order to receive signals, such as hormones, plant cells need receptors that specifically recognize each signal. Receptors, which are usually proteins, can be located on the surfaces or insides of cells.

Nucleic acids

Even if you haven't heard the term *nucleic acids* before, I'm sure you've heard of DNA, which is short for *deoxyribonucleic acid*. Nucleic acids like DNA are molecular specialists in information: The molecules are a chemical code that store information and can transfer it from one generation to the next.

Two types of nucleic acids are found in cells:

- ✔ **DNA stores the information that determines the structure and function of all cells on earth.** The structure and function of the cells lead to the traits of the organism, which is why people say that DNA determines your traits. People don't talk about plants much, but if they did, they'd say that DNA determines the traits of plants, too. You can think of DNA like the hard drive on a computer — it's the main place where information is stored. So, whether a plant becomes a mighty redwood or a tiny wildflower is ultimately encoded in the DNA of its cells. And just like the information in a computer, the information in DNA can be copied and transferred. When cells reproduce, they copy their DNA molecules and pass them on to the new cells.

- ✔ **RNA, or *ribonucleic acid*, is similar to DNA in structure, but more flexible in its functions.** Different types of RNA molecules perform different functions in cells: Some of them carry information around the cell, some of them help build proteins, and some of them control when proteins are made. In terms of information, RNAs are more like e-mails — they contain information, but they can travel around and cause things to happen.

Nucleic acids are made from *nucleotides*, which are complex molecules that consist of three parts (see Figure 2-6):

- **a 5-carbon sugar:** In RNA nucleotides, like the one shown in Figure 2-6, the sugar is *ribose*. In DNA nucleotides, the sugar is called *deoxyribose*. Deoxyribose looks just like ribose, except that it's missing one oxygen atom.
- **a phosphate group:** Phosphate groups contain a phosphorous atom surrounded by oxygen atoms. Some oxygen atoms have extra electrons, making them ionized and giving them a negative electrical charge. DNA and RNA molecules are negatively charged because of these phosphate groups.
- **a nitrogenous base:** Nitrogenous bases are ringed molecules that contain the element nitrogen. Five different nitrogenous bases are found in nucleotides: adenine (A), cytosine (C), guanine (G), thymine (T), and uracil (U). DNA nucleotides contain A, C, G, and T, while RNA nucleotides contain A, C, G, and U.

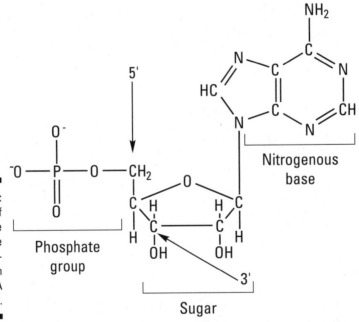

Figure 2-6: Structure of a nucleotide (adenosine monophosphate, an RNA nucleotide).

Cells make DNA and RNA molecules by forming covalent bonds between nucleotides (see Figure 2-7). The chains formed by this process are called *polynucleotide chains*.

DNA molecules contain two polynucleotide chains attached to each other by hydrogen bonds, while RNA molecules contain just one polynucleotide chain. The two polynucleotide strands of DNA attach to each other by hydrogen bonds between the bases A and T, and between the bases C and G, forming base pairs which look like the rungs of the ladder. The two strands twist around each other, forming the double helix of DNA.

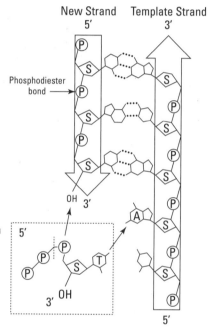

Figure 2-7: Synthesis of a polynucleotide chain.

The information code in DNA and RNA molecules depends upon the order of nitrogenous bases in the polynucleotide chain. Just like a computer stores information in bytes that consist of alternating 1's and 0's, or just like the 26 letters of the alphabet can write words, the pattern of chemical molecules A, C, G, and T spell out the information in DNA molecules, like the one shown in Figure 2-8. This information contains the instructions for building proteins and RNA molecules that determine the structure and function of cells. Similarly, the information in RNA molecules is spelled out in the order of the molecules A, C, G, and U.

 The genetic information of all cells is stored in molecules of DNA that are folded around proteins to form structures called *chromosomes*. The order, or *sequence*, of the four kinds of nucleotides within each chromosome spell out the instructions that determine the traits of the organism.

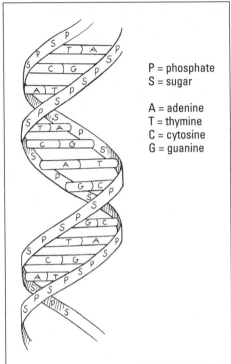

Figure 2-8: The twisted-ladder model of a DNA double helix.

P = phosphate
S = sugar

A = adenine
T = thymine
C = cytosine
G = guanine

Lipids

Lipids are molecules that don't mix with water, like fats, oils, and waxes. Cells, including those of plants, use lipids to create boundaries around and within cells. Lipids are also an excellent way to store energy and building materials for growth, and many plants, including olives, nuts, and even corn use oils as storage molecules.

All lipids have one structural feature in common — lots of carbon and hydrogen atoms bonded to each other with covalent bonds. These covalent bonds aren't polar at all, which means they are electrically neutral. (In other words, they're the opposite of the water molecules shown in Figure 2-3.) So, when lipids are mixed into water, all the water molecules stick to each other with hydrogen bonds but don't stick to the lipid molecules. It's like the water molecules don't want to play with the lipid molecules and push them all way. The result? The lipid molecules hang out together, floating on the surface of the water. Molecules like lipids that don't mix with water are called *hydrophobic*, which literally means water-fearing. In contrast, molecules that do mix with water are called *hydrophilic*, which means water-loving.

Hungry for hydrocarbons

The carbon-hydrogen bonds that make up lipid molecules store a great deal of usable energy. For example, you may have heard that fats have 9 calories per gram, while carbohydrates have 4 calories per gram. Gram for gram, fats store more than twice the amount of energy as carbs! But the human craving for hydrocarbons goes way beyond nutrition. Once humans discovered the potential of these molecules, we completely redesigned our way of life around them. The use of oil and gas to light our world extended our usable time into the hours of darkness. We hunted several species of whale almost to extinction for the oil we could extract from their bodies. Our use of hydrocarbons expanded when we learned to harness their power to run machines. The Industrial Revolution transformed our landscapes as we built factories and railroads, expanded our cities, and mined for gas, and oil (as well as coal, which is a carbon, but not a hydrocarbon). We could travel farther and build more than ever before.

Unfortunately, the side effects of this hydrocarbon habit are destruction of natural environments and pollution. One pollutant that's making itself felt today is the carbon dioxide (CO_2) that's released into the atmosphere when hydrocarbons are burned. Atmospheric CO_2 has increased steadily since the Industrial Revolution, and so have global temperatures. It turns out that CO_2 is a *greenhouse gas,* a gas that acts like a blanket and traps heat on the earth's surface. People today must simultaneously solve the problems brought on by our hydrocarbon fueled growth and figure out how we're going to replace our favorite energy source when it's gone! No one knows what the full extent of the consequences will be for the environmental changes we've caused and whether we can reverse the dangerous trends we've set in motion. On the energy front, we're already searching for solutions — making our machines more fuel efficient at the same time that we try to develop alternative fuel strategies (like wind and solar). Only time will tell if our harnessing of hydrocarbons was too much of a good thing.

Four types of lipids are especially important in plant cells:

- **Triglycerides (fats and oils):** *Triglycerides* store energy and building materials for growth. The structure of fats and oils is basically the same (see Figure 2-9): a 3-carbon molecule called glycerol forms the backbone to which three fatty acids attach. Most plants store oils, not fats.

  The difference between whether a triglyceride is a fat or an oil depends on how many *unsaturated* bonds it has between its carbon and hydrogen atoms. Unsaturated bonds result from two carbon atoms sharing two pairs of electrons from each other, forming a *double bond* like the one shown in the bottom fatty acid in Figure 2-9. Carbon atoms that are doing a double handshake with each other can't bond to as many hydrogen atoms, so the bonds are considered "not full" or unsaturated with hydrogen. *Saturated fats* contain lots of carbon atoms joined with single bonds, like the straight chains of fatty acids in Figure 2-9. Saturated fat molecules, like those in butter, can pack tightly together and are solid at room temperature. Unsaturated fats, like those in plant oils, have bent fatty acid chains, so they don't pack as tightly and are liquid at room temperature.

- **Phospholipids:** Cells build boundaries called *membranes* out of *phospholipids*. (To sneak a peek at phospholipids in membranes, go to "Customs: Plasma membranes," later in this chapter.) Phospholipids are similar in structure to triglycerides, but one fatty acid chain is swapped for a hydrophilic head group. So, phospholipids have a dual nature — they're hydrophilic at one end, and hydrophobic at the other.

- **Steroids:** Several plant hormones are *steroids,* lipid molecules made of four connected carbon rings. These hormones, called *brassinosteroids,* control many aspects of plant growth and development and trigger responses that protect plants from stress.

- **Waxes:** Many plants use *waxes* as a protective coating on the surfaces of leaves and other structures. Waxes help prevent water loss and can protect plants from insects and fungal pathogens. Carnivorous plants use waxes to make themselves slippery so that flies and other insects will slide to their doom! Waxes are diverse structurally, but the backbone of a wax is a long chain of carbon and hydrogen that is similar to a fatty acid. Next time you notice the gloss on a leaf, chances are you're looking at a plant wax.

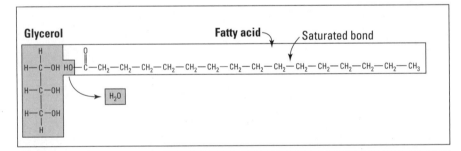

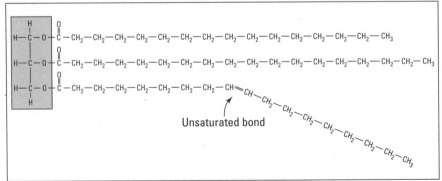

Figure 2-9: Saturated and unsaturated bonds in a typical triglyceride.

Entering the World of Cells

All living things, from tiny bacteria to enormous blue whales, are made of cells. Cells are the smallest things that have all the properties of life, including the ability to reproduce, respond to signals, grow, and transfer matter and energy with their environment. Figure 2-10 shows many of these functions and the structures that perform them in a plant cell.

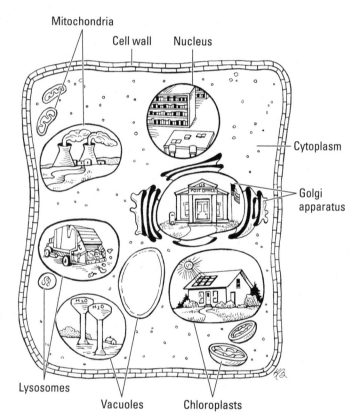

Figure 2-10: Plant cells perform all the functions of life.

Based on a comparison of fundamental cell chemistry and hereditary material, all cells on earth can be divided into three groups, called *domains*. You can think of these three domains as three main branches on the family tree of life on earth:

- **Eukarya:** Plants, animals, fungi, and protists
- **Bacteria:** Familiar, single-celled microorganisms, like the bacteria in your yogurt or the bacteria that cause human diseases

30 Part I: Plant Basics

✓ **Archaea:** Single-celled microorganisms that are found in all types of environments, but were first discovered in extreme environments like hot springs

In terms of cell structure, the cells of Bacteria and Archaea are very similar because they're both prokaryotic, while the cells of Eukarya are more complex because they're eukaryotic. (See the section "The Library: Storing Information in DNA," later in the chapter.)

The cells of organisms from all three domains have certain features in common. They all have a boundary that distinguishes the cell from the environment, contain DNA, and have the ability to make proteins. The next few sections present these common features of cells in more detail.

Customs: Plasma membrane

The barrier that separates the inside of the cell from its environment is called the *plasma membrane* (see Figure 2-11). The plasma membrane is made of two layers of phospholipids, forming a *phospholipid bilayer* with the hydrophilic heads of the phospholipids pointing outward and the hydrophobic tails of the phospholipids sandwiched in the middle. (For more details on phospholipids, check out the section "Lipids," earlier in this chapter.)

The job of the plasma membrane is to separate the chemical reactions occurring inside the cell from the chemicals outside the cell. Scientists say the plasma membrane is *selectively permeable,* which means it's choosy about what enters and exits the cell.

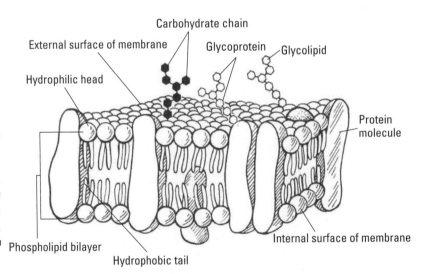

Figure 2-11: The fluid mosaic model of plasma membranes.

Proteins are also an important part of plasma membranes and help the plasma membrane do its job:

- **Proteins called *receptors* detect signals from the environment of the cell and relay the signal to the inside of the cell.** Scientists call the process of transferring a signal across a membrane *signal transduction*.
- ***Transport proteins* help control which molecules enter and exit the cell.**
  - Transport proteins called *channel proteins* form little tunnels in the membrane that can allow small molecules to pass quickly through the membrane.
  - Another type of transport protein, called *carrier proteins*, pick molecules up on one side of the membrane and then change their shape to deposit the molecules on the other side.

Scientists' refer to the structure of the plasma membrane as the *fluid mosaic model of the plasma membrane.* The membrane is a mosaic because it's made of different components, including phospholipids and proteins. It's fluid because it's flexible and molecules can move within it.

You can think of the plasma membrane like an international border that controls what enters and leaves the country. The proteins that regulate movement across the membrane are the customs officials.

Just inside the plasma membrane is the *cytoplasm* of the cell, the fluid material that contains all the molecules and structures of the cell. The cytoplasm is a busy place, filled with chemical reactions and moving materials.

Botanists call everything inside a plant cell, besides the nucleus and the cell wall, the *protoplast*.

Think of the cytoplasm of a cell like downtown in a busy city. The buildings in downtown are like the organelles of the cell, each specialized for a different function. And the cars speeding through downtown are like the materials that constantly moving around in cells.

The library: Storing information in DNA

Just like people store information in libraries, all cells store information in deoxyribonucleic acid. (For the scoop on DNA, see the earlier section "Nucleic acids.") One major difference between prokaryotic and eukaryotic cells, however, is that they package their DNA differently:

- *Eukaryotic cells* separate their DNA from the cytoplasm in the nucleus of the cell by surrounding the DNA with a sphere of membrane called the *nuclear membrane*. Like the plasma membrane, the nuclear membrane is mostly made of phospholipids and proteins. However, the nuclear membrane is actually a double membrane made up of two phospholipid bilayers. Materials and structures can travel in and out of the nucleus through little protein tunnels in the nuclear membrane called *nuclear pores*.
- *Prokaryotic cells* locate their DNA directly within the cytoplasm in a region of the cell called the *nucleoid*.

Factories: Ribosomes

Ribosomes are small structures in cells that help build proteins. Because proteins are very important workers in cells, all cells need ribosomes.

The ribosomes in all types of cells have a very similar structure:

- Ribosomes are made of two types of molecules: a special type of RNA, called *ribosomal RNA* (rRNA), and proteins
- The rRNA and proteins of ribosomes are twisted together to form two separate pieces: the *large subunit* and the *small subunit*. These subunits are built separately from each other and come together to form a completed ribosome when a cell begins to make a protein.

Think of ribosomes as little factories where proteins are built.

To make a protein, cells complete two processes:

- **Transcription:** In the nucleus, cells copy the instructions for the protein from the DNA into a new molecule, called *messenger RNA* (mRNA). The mRNA leaves the nucleus and carries the instructions to the ribosomes out in the cytoplasm of the cell.
- **Translation:** Ribosomes organize the mRNA and other molecules that are needed to build proteins and help to put proteins together.

Exploring Plant Cells

Plants, like other eukaryotes, have cells that are highly organized, with lots of compartments for different functions. In addition to the nuclear membrane that walls off the DNA, other membranes form structures called *organelles* that perform specialized tasks for the cell. Plants and animals are both eukaryotes, so their cells have lots in common, which you can see if you compare the two types of cells shown in Figures 2-12 and 2-13. In fact, most people would be very surprised to learn how much they have in common with plants!

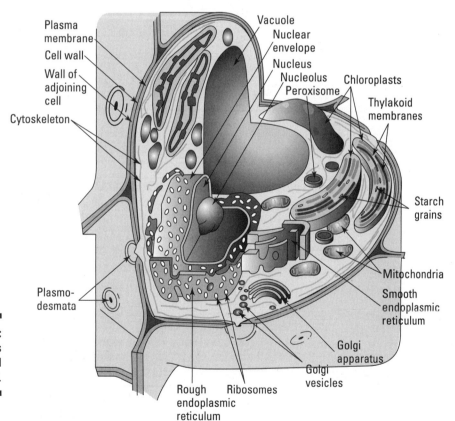

Figure 2-12: Structures in a typical plant cell.

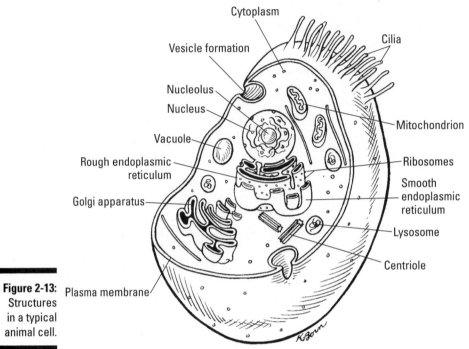

Figure 2-13: Structures in a typical animal cell.

Almost all eukaryotic cells, including those of plants and animals, contain the following structures:

- Plasma membrane
- Ribosomes
- Cytoplasm
- Nucleus
- Endoplasmic reticulum
- Golgi apparatus (sometimes called *dictyosomes* in plant cells)
- Mitochondria
- Cytoskeleton
- Vesicles, vacuoles, and lysosomes

In addition, plants have four structures, some of which are found in other eukaryotes, but which are not found in animal cells):

- Cell wall
- Plastids, including chloroplasts
- Large central vacuole
- Plasmodesmata

The following sections cover the cellular structures that are common to all eukaryotes, as well as those that are unique to plants.

Post office: The endomembrane system

The *endomembrane system* is a large network of membranes that helps construct proteins and lipids, packages them up, and then ships them where they need to go. Think of the endomembrane system as the Post Office.

All the components of the endomembrane system are made of membranes that contain phospholipids and proteins:

- The *endoplasmic reticulum,* which scientists call the ER, is a set of folded membranes that begins at the nucleus and extends into the cytoplasm. The ER begins with the outer membrane of the nuclear envelope and then twists back and forth like switchbacks on a steep mountain trail. There are two types of endoplasmic reticulum:
    - *Rough endoplasmic reticulum* (RER) is called "rough" because it is studded with ribosomes. Ribosomes that begin to make a protein that has a special destination, like a particular organelle or membrane, will attach themselves to the rough endoplasmic reticulum while they make the protein. After the protein is made, it is packed in a membrane sphere called a *vesicle,* and then shipped to the Golgi apparatus.
    - *Smooth endoplasmic reticulum* (SER) does not have any attached ribosomes. It makes lipids; for example, phospholipids for cell membranes.
- The *Golgi apparatus* is a stack of flattened membrane sacs that looks a little bit like a stack of pancakes plus assorted vesicles that are traveling to and from the Golgi. Proteins travel in vesicles to the Golgi from the RER. When proteins arrive at the Golgi, they're modified, marked with chemical tags, and sorted. The proteins leave the Golgi in vesicles and are shipped to their proper destination. Scientists who study plants have their own name for the stack of pancakes part of the Golgi — they call it a *dictyosome.* But, dictyosomes and Golgi apparatus are really pretty much the same thing, and Golgi apparatus is a more widely used term.

- *Vesicles* are little spheres of membrane in the cell and come in several types:

  - *Transport vesicles* carry things around the cell. They're like the Priority Mail boxes that you put your letters in. They travel from the ER to the Golgi and then to the plasma membrane or organelles to bring molecules where they need to go.

  - *Lysosomes* are the garbage disposals of the cell. They contain digestive enzymes that can break down molecules, organelles, and even bacterial cells.

  - *Secretory vesicles* bring materials to the plasma membrane so they can be released, or secreted, from the cell.

- *Peroxisomes* are small organelles encircled by a single membrane. Often, they're involved in breaking down lipids like fatty acids. In plants cells, *glyoxisomes*, a special kind of peroxisome, help convert stored oils into molecules that can be easily broken down for energy.

- *Vacuoles* are similar to vesicles in structure — they're basically spheres of membrane, but they have different functions. The majority of space inside most plant cells — up to 95 percent of the cell volume! — is taken up by a *large, central vacuole*. The vacuole is separated by a membrane called the *tonoplast*. Plants use their central vacuole as a place to dump wastes and store materials.

Think of a plant cell like a water balloon inside a shoe box. The shoe box represents the cell wall that surrounds the plant cell, and the water balloon is the large, central vacuole. The latex of the balloon is like the membrane around the vacuole. All of a plant's cytoplasm is squeezed between this membrane and the cell wall.

Scaffolding and railroad tracks: The cytoskeleton

A network of protein cables called the *cytoskeleton* runs throughout cells, reinforcing their structure and enabling the movement of cells and materials.

Cytoskeletal proteins have three main functions in cells:

- **Scaffolding:** Cytoskeletal proteins underlie membranes, giving them shape and support, much like scaffolding can support a building.

- **Movement of materials within cells:** Cytoskeletal proteins run like railroad tracks through cells, enabling the movement of vesicles and organelles along the proteins.

- **Movement of cells:** Cytoskeletal proteins are found within whip-like cellular extensions called *cilia* and *eukaryotic flagella*. Cells like sperm cells (including the sperm cells made by some plants!) swim by flicking their flagella.

Three types of cytoskeletal proteins reinforce and organize eukaryotic cells:

- **Microfilaments** are made of the protein *actin*. In plant cells, microfilaments are involved in cell division and expansion and act as railroad tracks for vesicles and organelles, helping to circulate the cytoplasm in a process called *cytoplasmic streaming*. In animal cells, microfilaments make muscle cells contract and help pinch cells in two during cell division.
- **Microtubules** are made of the protein *tubulin*. Microtubules are the proteins inside of cilia and flagella. Microtubules also move chromosomes during cell division, and act as railroad tracks for the movement of vesicles and some organelles. Microtubules help organize the formation of the plant cell wall and thus determine the development of plant shape.
- **Intermediate filaments** are made of various proteins that tend to be strong proteins that help reinforce cellular structures. For example, the protein *lamin* strengthens the nuclear membrane, and the tough protein *keratin* helps reinforce your skin cells and some plant cells, like those in the roots that have to push through the soil.

Solar-powered batteries: Chloroplasts

Most of the food on planet earth is made by organisms that do *photosynthesis*. During photosynthesis, cells transform light energy from the sun into chemical energy stored in food. Photosynthesis occurs in the cells of plants, algae, and bacteria. The eukaryotic cells of plants and algae contain a special compartment for photosynthesis, a green organelle called a *chloroplast* (see Figure 2-14).

Chloroplasts are green because they contain *chlorophyll,* a green pigment that can absorb sunlight. During photosynthesis, the energy from sunlight is used to combine the atoms from carbon dioxide and water to produce sugars, from which all types of food molecules can be made. (To make proteins, cells also require a source of nitrogen.)

You can think of chloroplasts as solar-powered batteries because they store energy from the sun as chemical energy in food.

Chloroplasts are bordered by two membranes, an *inner membrane* and an *outer membrane*, both of which are phospholipid bilayers. In addition, they have little sacs of membranes called *thylakoids* stacked up in towers called *grana*. You can see each of these structures in Figure 2-14.

The membranes of the chloroplasts divide them into several different compartments that allow the different reactions of photosynthesis to be separated from each other. (For more details on photosynthesis, see Chapter 7.)

- The *intermembrane space* is between the inner and outer membranes.
- The *stroma* is the fluid-filled space surrounded by the inner membrane.
- The *interior of each thylakoid* is another fluid-filled space.

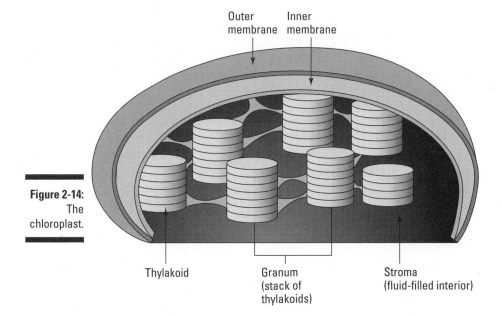

Figure 2-14: The chloroplast.

Chloroplasts are a type of *plastid,* self-replicating organelles that are surrounded by a double membrane and that perform important chemical processes in plant cells. The function of chloroplasts is photosynthesis and they're found in the green parts of plants.

Plant structures that specialize for other functions may contain different types of plastids:

- *Proplastids*, also called eoplasts, are small colorless plastids that give rise to all the types of plastids. Proplastids are found in seedlings and plant tissues that haven't yet specialized for a certain function.
- *Leucoplasts* are colorless plastids that specialize in the synthesis and storage of important molecules. Three different types of leucoplasts are found in plants:
  - *Amyloplasts* make and store starch. They're typically found in roots and other plant tissues that store starch. For example, potatoes and other tubers are loaded with amyloplasts.
  - *Elaioplasts*, also called oleoplasts, make and store oil. Elaioplasts are found in plant tissues that provide support to developing pollen grains.
  - *Proteinoplasts,* also called proteoplasts or aleuroplasts, make and store proteins. Proteinoplasts are found in many seeds, such as peanuts.
- *Chromoplasts* are red, orange, and yellow-colored plastids that contain pigments called *carotenoids*. Chromoplasts are found in flowers, fruits, and leaves that are turning colors in the fall. Chromoplasts can form from chloroplasts and amyloplasts as well as from proplastids.
- *Etioplasts* are pale, yellowish plastids found in plant tissues growing in the dark. When plants are exposed to light, etioplasts develop into chloroplasts.

Powerplants: Mitochondria

Photosynthetic organisms make food, but all organisms break food down in order to obtain matter and energy for growth. In eukaryotes, most of the energy release from food occurs through a process called *cellular respiration* inside an organelle called the *mitochondrion* (see Figure 2-15). During cellular respiration, cells transfer the energy stored in food into a molecule called *adenosine triphosphate* (ATP), which cells can easily use to provide energy for cellular reactions. (To get more information on ATP, check out Chapter 6. For more details on cellular respiration, see Chapter 8.)

A very common — but wrong! — idea is that plants have chloroplasts instead of mitochondria. The right idea is that plants have both! Think about it — it wouldn't do plants very much good to make food if they couldn't also break it down! When plants make food, they store matter and energy for later. When they need the matter and energy, they use their mitochondria to break the food down.

Like chloroplasts, mitochondria are surrounded by two membranes: the *outer membrane* and the *inner membrane*. In addition, the inner membrane is folded back and forth, creating folds called *cristae* that are needed during cellular respiration to transfer energy to ATP.

The membranes of the mitochondrion divide it into two different spaces (see Figure 2-15):

- The *intermembrane space* is between the two membranes of the mitochondrion.
- The *matrix* is the fluid-filled space inside of the inner membrane.

You can think of mitochondria as the power plants of the cell because they transform energy from one form (chemical energy in food) to a form that is more useful for cells (chemical energy in ATP).

Rebar and concrete: Cell walls and extracellular matrices

Most cells have additional layers outside of the plasma membrane that provide additional strength to cells and may attach cells together in organisms like plants that are made of more than one cell. Many of these extracellular layers use a strategy that's a lot like putting rebar into concrete in order to make a wall stronger — in cells, long cables of carbohydrates or proteins act like the rebar and a sticky matrix takes the place of the concrete.

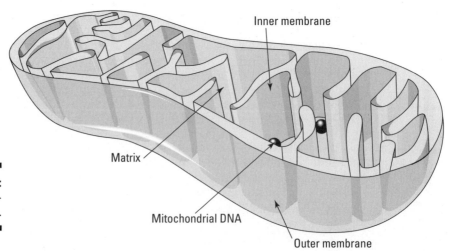

Figure 2-15: The mitochondrion.

Two types of strengthening layers are formed outside of cells:

- *Cell walls* are semi-rigid reinforcing layers that help protect the cell. Many eukaryotes, including plants (see Figure 2-16), fungi, and some protists, make cell walls, as do most bacteria.

- The *extracellular matrix* (ECM) is a flexible layer formed around animal cells. The ECM is made of long proteins, such as *collagen*, embedded in a polysaccharide gel.

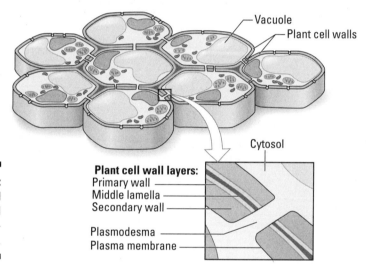

Figure 2-16: Plant cell walls and plasmodesmata.

The plant cell wall can have several different layers, which you can see in detail in Figure 2-16. Plants make their cell walls from the outside in – outside layers first, followed by the inside layers:

- The *middle lamella* is the outermost layer that helps stick plant cells together. It forms first when plant cells divide. The middle lamella is a thin, flexible layer made mostly of sticky polysaccharides called *pectins*.

- Plants produce the *primary cell wall* under the middle lamella before and during active growth. The polysaccharide *cellulose* forms strengthening fibers that are embedded in a sticky matrix of pectin and other polysaccharides. The primary cell wall may be thicker than the middle lamella.

- Woody plants produce a *secondary cell wall* and deposit it under the primary cell wall when growth is complete. The secondary cell wall is mostly composed of cellulose and a strong, complex chemical called *lignin*. Lignin is a large strong polymer that acts like superglue over the primary cell wall. Secondary cell walls are the thickest layer of the cell wall.

Plant cells are protected and stuck together by their cell walls, but they aren't completely walled off from each other. Little tunnels, called *plasmodesmata,* pass through the cell walls, creating connections between neighboring cells. These connections are essential to the movement of materials and communicating molecules between cells.

Chapter 3
Identifying Plant Tissues

In This Chapter
- Making more cells with meristems
- Comparing simple and complex tissues
- Discovering the types of cells found in plant tissues

Plants grow from their tips as cells in their meristems divide to produce new cells. Cells differentiate, becoming specialized to perform specific functions, combining with other cells to form unique types of tissues in plants. This chapter presents the different types of tissues and tissue systems that are found in plants.

Combining Cells to Form Tissues

Plants grow as cells divide to produce new cells in areas of the plant called *meristems*. After new cells are produced, they specialize for certain functions, enlarging and changing their structure to match their function. Just as in animals, plant cells are organized into *tissues,* which are collections of cells that work together to perform a function. Tissues are combined together to form *tissue systems.* Tissue systems then make up the *organs* of the plant, which are organized structures such as stems, roots, and leaves that perform specific functions for the whole plant.

Plants have four main types of tissues systems:

- The *meristematic tissue system* contains cells that are actively dividing to produce new plant tissue.
- The *ground tissue system* contains tissues that make up the bulk of the plant, including those that do photosynthesis, support the plant, store food, and repair damage.

- The *vascular tissue system* contains tissues that conduct water, minerals, and sugars throughout the plant.
- The *dermal tissue system* contains tissues that protect the plant and prevent the loss of water from tissues.

Growing with meristems

You've probably heard of *stem cells,* cells in animals that have the ability to become any type of cell in the body. Plants have something similar in their *meristematic tissues,* tissues where cells are actively multiplying to provide new cells for plant growth. New cells produced in the *meristem* are all-purpose cells, just like animal stem cells, because they can become any type of plant cell. Cells in the meristem even look generic — they're small, box-shaped cells, with small vacuoles (Figure 3-1). As the cells enlarge, they begin to *differentiate,* becoming specialized for a certain function. Their appearance changes as they change their structure to match their new function and they become recognizable as a certain type of plant cell.

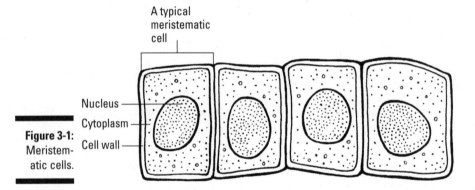

Figure 3-1: Meristematic cells.

A plant has two organ systems:

- The *shoot system,* the parts of the plant that grow above ground and support leaves, flowers, and fruits
- The *root system,* the underground part of the plant that absorbs water and minerals

Figure 3-2 shows the locations of meristems in the shoots and roots. Plants have four types of meristems:

- Apical meristems
- Axillary buds
- Lateral meristems
- Intercalary meristems

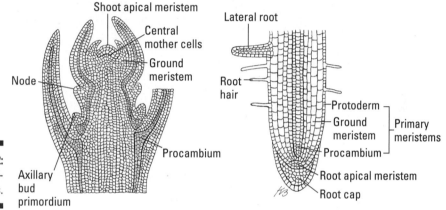

Figure 3-2: Plant meristems.

Apical meristems

Plants get taller, and roots get longer, from their tips.

Apical meristems are areas of rapidly dividing cells found at the tips of roots and shoots. (*Apical* means from the apex, or from the tip.)

The apical meristems of roots and shoots divide to produce new cells, which then elongate, making roots get longer and shoots get taller. (For more details on plant cell division, head to Chapter 11.)

Once cells elongate and differentiate into a particular cell type, they usually can't divide any more due to their rigid cell walls. Meristematic cells, however, continue to divide, enabling growth throughout the life of the plant. The ability to grow throughout life is called *indeterminate growth*. Animals, including people, usually demonstrate *determinate growth,* growth that stops once a structure is fully formed. To understand the differences between these types of growth, compare the height of a tree to that of a person. Trees keep growing taller until they die, whereas people stop getting taller once they reach maturity (which is probably a good thing!).

If you hung a swing from the branch of a tree when the tree was young, the swing wouldn't rise up into the air as the tree got taller. That's because the tree is getting taller from the top, as the trunk grows longer from its tip.

Axillary buds

As stems elongate, they produce small buds, called *axillary buds,* in the angle between the stem and the petiole of each leaf. Axillary buds may remain dormant for a time, until the apical meristem of the main shoot has grown for a while and moved away from the axillary buds. After the apical meristem is far enough away, the meristems of the axillary buds begin producing new cells, and a branch begins to grow — from its tip, of course, just like any other shoot. Branches are just shoots that emerge from points, called *nodes,* along the main shoot. (I discuss the plant hormones that control the dormancy of the axillary buds in Chapter 10.)

If you've ever pinched off the tip of a plant in order to encourage more branching, the reason it worked was because you helped wake up those axillary buds by removing the apical meristem of the main shoot.

Lateral meristems

Lateral meristems are thin cylinders of meristematic tissue that form in mature regions of shoots in roots in many plants, especially those that produce woody tissue. Lateral meristems divide to produce *secondary growth,* growth that increases the diameter of a shoot or root.

If you tied a ribbon around the trunk of a young tree, the ribbon would eventually break as the tree got thicker. Lateral meristems produce the cells that make the tree get thicker.

Plants have two types of lateral meristems:

- The *vascular cambium* produces new vascular tissue.
- The *cork cambium* divides to produce *cork,* a tissue found in the outer bark of woody plants.

If you think of the vascular cambium as a straw of tissue inserted into a stem or root, then the cork cambium is a slightly larger straw that surrounds the first — to visualize this more clearly, refer to Figure 3-2.

Intercalary meristems

Intercalary meristems are regions of meristematic tissue that occur at intervals along the stems of grasses and related plants. When cells of the intercalary meristem divide, they add new cells to the stem, increasing its length.

Chapter 3: Identifying Plant Tissues 47

 You can mow your grass, chopping off all the plants' apical meristems. For plants that don't have intercalary meristems, the loss of the apical meristem greatly impacts their ability to grow taller. Not so for grasses! When you cut off their apical meristems, they continue to get longer from the division of their intercalary meristems. The result? You have to mow again!

Building plant bodies with ground tissue

Most of the primary growth of plants consists of ground tissue. Because it occurs everywhere in plants, it's often interspersed with other tissues. Figure 3-3 shows the three types of simple tissues that make up the ground tissue system:

- Parenchyma
- Collenchyma
- Schlerenchyma

 Simple tissues typically consist of only one type of cell, whereas *complex tissues* contain more than one type of cell.

Parenchyma

Parenchyma cells are your general, all-purpose type of plant cell. They're the most abundant type of cell in a plant and can perform several different functions, such as metabolism and storage of different materials. For example, parenchyma cells in leaves are full of chloroplasts so they can do lots of photosynthesis, whereas parenchyma cells in a storage organ, like a potato tuber, store lots of starch grains in amyloplasts. Parenchyma cells may also contain crystals, oils, or other types of plastids. (For more on plastids, see Chapter 2). *Secretory cells* are cells derived from parenchyma cells. These cells produce special substances inside the plant body and then store the materials in channels before releasing them to the outside of the cell, or even to the outside of the plant body.

Parenchyma has two key characteristics that can help you distinguish it from other tissues:

- **Thin walls:** Parenchyma cells have a primary cell wall of cellulose and remain relatively flexible. Depending how they're pushed together in a tissue, they can have slightly different shapes, but they're often block-shaped or rounded.
- **Alive at maturity:** Parenchyma cells remain alive and still have the potential to divide when they're fully developed. They can live for a really long time — up to 100 years in some cactus plants!

TECHNICAL STUFF Parenchyma cells that are full of chloroplasts are called *chlorenchyma*. Parenchyma cells with lots of interconnected air spaces, like those in aquatic plants, are called *aerenchyma*.

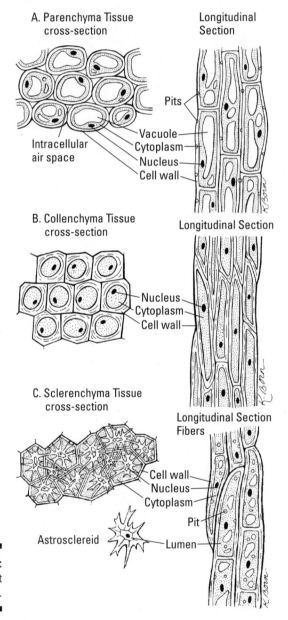

Figure 3-3: Simple plant tissues.

Collenchyma

Collenchyma cells are very similar to parenchyma cells, with the exception of having thickened cell walls. The thickened yet flexible cell walls enable collenchyma tissue to support both growing and mature plant organs.

If you want to get a real feel for collenchyma, eat a stalk of celery. The long flexible strings that tend to get stuck in your teeth contain collenchyma. They're strong, yet flexible, providing bendable support to the celery stalk.

Collenchyma has three key characteristics that can help you distinguish it from other tissues:

- **Thickened cell walls:** The cell walls of collenchyma cells are thickened unevenly and may be thickest at the corners (refer to Figure 3-3). The cell walls of collenchyma are primary cell walls, so they don't contain lignin, which helps keep the tissue from becoming rigid.
- **Location just under the epidermis:** Parenchyma tissues may be located throughout the plant, but collenchyma is typically found just under the epidermis, which is the outermost layer of cells on the plant.
- **Alive at maturity:** Like parenchyma cells, collenchyma cells are functioning cells that can live a long time.

Sclerenchyma

Sclerenchyma cells have thick cell walls that contain lignin, which makes them very tough and supportive. Sclerenchyma protects plant organs from damage due to stresses, such as bending and stretching.

If you want to check out some sclerenchyma, you could eat a pear. The gritty texture of a pear is due to a type of sclerenchyma cell called *stone cells* that are embedded in the fruit. Or try testing the hardness of a peach or plum pit — the outer layer of the pit is hardened with lots of sclerenchyma cells.

Sclerenchyma has three key characteristics that can help you distinguish it from other tissues:

- **Thick cell walls:** Sclerenchyma cells have thick secondary cell walls that contain lignin and are laid down inside the primary cell wall. To see how thick the cell walls can get, see the stone cells in Figure 3-3.
- **Dead at maturity:** Sclerenchyma cells die after they become fully mature. The cytoplasm breaks down, leaving just the cell wall behind.
- **Location throughout plant:** Sclerenchyma can be located in many different tissues (unlike collenchyma which is typically located just under the epidermis).

Fashionable fibers

Do you have any clothing or household items made out of linen? If you do, then you know first hand one of the uses that people have found for sclerenchyma. Linen is made from the flax plant (*Linum usitatissimum*). Schlerenchyma fibers are removed from the stem, spun into thread, and then woven into linen cloth. Before cotton was mass produced during the Industrial Revolution in the 19th century, linen was the fabric of choice for societies from the ancient Egyptians to the British colonials. Linen was such an important fabric that today people use the phrase "fine linens" to refer any high-quality house fabrics. You can read more about linen and other textiles made from plants in Chapter 21.

Sclerenchyma cells are divided into two groups: *sclereids* and *fibers*. Sclereids are shorter than fibers — usually about as long as they are wide. Two types of sclereids, stone cells and astrosclereids, are shown in Figure 3-3. Fibers, which are also shown in Figure 3-3, are much longer than they are wide and are tapered at the ends. Fibers usually occur in strands, with the individual fibers overlapping slightly.

Protecting plants with dermal tissue. Dermal tissues cover plant surfaces or create barriers that protect the plant. Scientists include three tissues in the dermal tissue system:

- Epidermis
- Endodermis
- Periderm

Epidermis

Plants are covered by a single layer of cells, called the *epidermis*, that separates the inside of the plant from its environment. The *epidermis* is made mostly of parenchyma-like cells that are alive at maturity.

The epidermis has three primary functions:

- **Prevent water loss:** Most epidermal cells secrete a waxy material called *cutin*, which forms a waxy layer called the *cuticle* on the outside of the plant. The cuticle provides water-proofing to the plant tissues so that their water doesn't just evaporate into the air. Epidermal cells are also packed tightly together with very little space between the cells so that water can't evaporate out from between them.

The epidermis of the shoot system may also produce little hair-like projections called *trichomes*. Trichomes have many functions depending on the plant that produces them. In some plants, trichomes help prevent water loss by creating a barrier to air movement along the surface of the plant that could speed up evaporation. Plants may also use trichomes for defense and seed dispersal. Some trichomes are glandular and contain secretory cells that produce materials like oils.

- **Protect plant from fungi and other attackers:** The waxy cuticle and closely spaced cells of the epidermis help prevent fungi, bacteria, and insects from being able to access the plant tissues. The wax makes the cells harder to penetrate, and the tightly packed cells prevent attackers from slipping into the interior of the plant through gaps between the cells.

- **Allow plant to exchange materials with the environment:** Plants have to strike a balance between preventing water loss and being able to get what they need from their surroundings. In order to live and grow, plants need to be able to move materials through the epidermis:

  - *Plants do gas exchange with the atmosphere:* Plants need to take carbon dioxide (CO_2) from the air in order to do photosynthesis (see Chapter 7), and they release oxygen gas (O_2) as waste. The waxy cuticle is great for stopping water loss to the air, but it also prevents gas exchange. So, plants need openings in the epidermis (sort of like you need nostrils in your nose). These openings are called *stomata* or *stomates* (singular = *stoma*). Plants control whether their stomata are open or shut with special cells called guard cells.

  - *Plants absorb water and minerals from the soil:* Roots absorb water and minerals from the soil through their surface cells. To help create more surface for water absorption, root epidermis produces little projections called *root hairs*. Plants like orchids that grow up in the trees produce aerial roots called *velamen*. The epidermis of velamen is several cells thick and modified so that it's very absorptive, allowing the plant to capture water and minerals from the air.

Endodermis

The endodermis is a layer of cells between the cortex and vascular tissue in roots, as well as some stems and leaves. The endodermis is very important to the overall water balance of plants because it helps regulate the uptake of water from the soil and helps prevent the loss of water from plants.

Periderm

The periderm is the outermost tissue in woody plants, forming the outer bark. Periderm is a complex tissue, consisting mostly of cork cells and parenchyma cells:

- ✓ **The cork cells of the periderm help protect the stem.** The stems of woody plants get thicker as the cork cambium and vascular cambium begin to divide. As the stems get thicker, the original epidermis cracks and falls off. A new protective layer forms from the box-shaped *cork cells* produced by the cork cambium. The cork cells produce a waxy substance called *suberin,* which helps protect the stem of the plant from water loss and attack from fungi and other pests. Once the cork cells reach maturity, they die.

- ✓ **The parenchyma cells of the periderm allow gas exchange with the environment.** The cork cambium also produces clusters of parenchyma cells, forming pockets of tissue called *lenticels*. The parenchyma cells aren't waxed, so they can still exchange gasses like carbon dioxide (CO_2) and oxygen (O_2) with the environment.

You can see lenticels on the surfaces of some darker woods, like horse chestnut, birch, and cherry. Lenticels look like little eye-shaped lighter areas on the wood.

Transporting materials with vascular tissue

The vascular tissue system acts as the plant's plumbing, moving water and sugars throughout the plant. The vascular tissue system contains two types of complex tissues:

- ✓ **Xylem:** Transports water and minerals throughout the plant body.
- ✓ **Phloem:** Transports sugars throughout the plant body.

Xylem

Xylem is like the plumbing system of a plant, transporting water and minerals vertically through the plant from the roots to the leaves.

Xylem is a complex tissue consisting of several cell types, including the distinctive tracheids and vessel elements shown in Figure 3-4. In addition to transporting water and minerals, the secondary cell walls in xylem helps strengthen and support plant stems.

Xylem contains as many as five types of cells:

- ✓ **Tracheids** are long thin cells with pointed ends that conduct water vertically. Tracheids line up in columns like pipes by overlapping their tapered ends. Tracheids die when they reach maturity, so water is conducted through the tubes made up of tracheid cell walls.

  Wherever two ends join, small holes in the cell wall called *pits* line up to allow water to flow from one tracheid to another. Pits always occur in pairs so that a pair of pits lines up on either side of the middle *lamella,* or center layer, of the cell wall. (See Chapter 2 for more details on plant cell walls.)

✔ *Vessel elements* are barrel-shaped cells with open ends that conduct water vertically. Like tracheids, vessel elements line up end to end forming columns, called *vessels,* that conduct water. Some vessel elements have completely open ends, while others have narrow strips of cell wall material that partially cover the ends. Vessel elements, like tracheids, die when they reach maturity.

✔ *Ray cells* are long-lived parenchyma cells that extend laterally like the spokes of a wheel from the center of a woody stem out towards the exterior of the stem. Ray cells are alive at maturity, and they transport materials horizontally from the center of a stem towards the exterior.

✔ *Xylem fibers* are long, thin sclerenchyma cells that run parallel to the vessel elements. Xylem fibers help strengthen and support the xylem.

✔ *Xylem parenchyma cells* are living parenchyma cells distributed among the vessels and tracheids. Xylem parenchyma cells store water and nutrients.

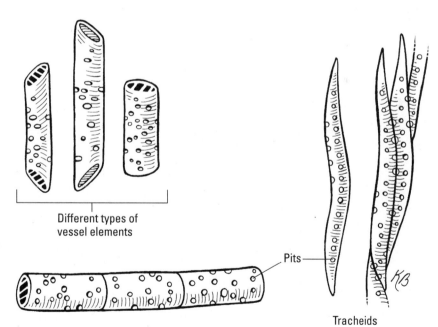

Figure 3-4: Types of water conducting cells.

When scientists look at plant fossils, they find evidence of tracheids in the oldest plant fossils, with vessels found in more recent fossils. Because of this evidence, scientists believe that tracheids evolved first and that vessels evolved later. Today, tracheids are found more commonly in gymnosperms, including conifers, while vessels and tracheids are found in flowering plants. The trend in vessel elements over time seems to move away from a tracheid-like shape of thin and tapered, to a more open, barrel-like shape. The ends of vessel elements also seem to evolve from partially covered to completely open.

Phloem

Phloem acts like the circulatory system of a plant, bringing sugar from the leaves to all parts of the plant body. Phloem is a complex, living tissue at maturity that contains several types of cells, including the two most common types, sieve tube elements and companion cells, shown in Figure 3-5. Like the xylem, phloem also contains rays, fibers, and parenchyma cells.

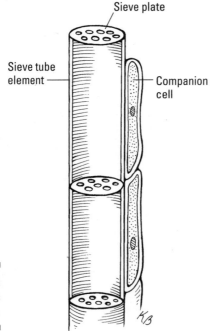

Figure 3-5: Components of phloem.

Phloem contains as many as five types of cells:

- **Sieve-tube elements** are long, straw-like cells that link themselves end to end to form *sieve tubes* that conduct sugar throughout the plant. The ends of sieve-tube elements have lots of holes in the cell wall, forming a perforated plate called a *sieve plate* that allows the sugar solution to pass from one cell to the next in the sieve tube. Sieve tube elements are alive at maturity, but they have no nucleus of their own, relying instead upon their adjacent companion cells.

- **Companion cells** are living, nucleated cells that occur adjacent to sieve-tube elements. Companion cells regulate the sieve-tube elements and help them maintain themselves.

- **Phloem ray cells** extend laterally through stems allowing horizontal transport of materials. Phloem rays and xylem rays form from the same cells and are very similar in structure and function.

- **Phloem fibers** are long, thin sclerenchyma cells that run parallel to the sieve-tube elements. Phloem fibers help strengthen and support the phloem.

- **Phloem parenchyma cells** are living parenchyma cells located between the sieve tubes. These cells store water and food.

Chapter 4
Vegetative Structures: Stems, Roots, and Leaves

In This Chapter
▶ Defining vegetative plant organs
▶ Distinguishing between primary and secondary growth
▶ Examining the structure and function of stems, roots, and leaves

*P*lants produce vegetative and reproductive organs. Vegetative organs, such as stems, roots, and leaves, help plants make and use food, obtain water and minerals, and store resources, like food and water. Differences in the structure of stems and roots occur between two major groups of flowering plants, the monocots and the dicots, and also between herbaceous plants that show only primary growth and woody plants that show secondary growth.

In this chapter, I present the basic structure and function of vegetative plant organs and introduce you to the diversity of these organs among different plant groups.

Getting Organized into Plant Organs

You're no doubt aware of your own organs like your heart, kidney, or lungs. You may not have thought of plants as having organs, too, but they do!

Plant *organs* are groups of several types of tissues that together perform a particular function.

Vegetative organs — stems roots and leaves — make and use food, absorb water and minerals, transport materials throughout the plant, and store food. Reproductive organs, such as flowers and cones, participate in sexual reproduction. (For more info on reproductive organs, see Chapter 5. For the details on sexual reproduction, see Chapter 11.)

Within the plant body, three major organs perform all the vegetative functions necessary for life:

- **Stems** support the leaves and transport water, minerals, and sugars throughout the plant.
- **Roots** anchor the plant in the ground and absorb water and minerals from the soil.
- **Leaves** make sugars through the process of photosynthesis (see Chapter 7).

Getting Taller with Stems

Anytime you use something made of wood, you're using something made from the stem of a plant. *Stems* are linear structures with attached leaves that provide support and transport of water and nutrients to the plant body. Most stems grow upward, helping to raise plant structures, such as leaves, off the ground. Some plants also use their stems for photosynthesis or for food and water storage.

You can see the parts and organization of stems fairly easily on a woody stem, like the one shown in Figure 4-1:

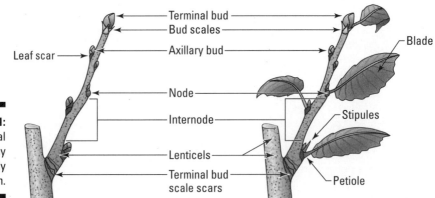

Figure 4-1: External anatomy of a woody stem.

- **The apical meristem is located in a bud, called a *terminal bud*, at the tip of the stem.** New growth in length occurs because of cell division in the apical meristem. Along with the apical meristem, terminal buds may contain *leaf primordia,* fully formed, tiny embryo leaves that are ready to expand and grow when the bud opens. Buds are usually protected by hard, leaf-like structures called *bud scales*. When the terminal bud begins to grow, the bud scales fall off, leaving a ring of scars that go 360 degrees all around the stem.

For deciduous trees growing in temperate climates, you can tell how many years a woody branch has been growing by counting the number of rings of bud scale scars.

- **Leaves grow from the stem at *nodes*.** On branches from plants that drop their leaves, called *deciduous plants, leaf scars* show where the leaves used to be attached. If you look closely at leaf scars, you can see little circular marks within them that show where the vascular tissue from the stem ran out into the leaf. These vascular scars are called *bundle scars* because they mark the places where bundles of vascular tissue were located. The spaces between nodes on the stem are called *internodes*.

- **Lateral meristems are located in *axillary buds* that are tucked in the angles, or *axils*, between the leaves and the stem.** When an axillary bud begins to grow, its lateral meristems function just like the apical meristem producing new growth through cell division. Axillary buds may grow into new branches, or they may produce flowers. Just like terminal buds, axillary buds are usually protected by bud scales.

All plants grow by getting taller as the apical meristem produces new cells at the tip of the stem. Additionally, the stems of some plants that grow for more than one year will grow thicker over time. So, two types of growth occur in stems:

- **Primary growth increases the length of the stem.** Primary growth results from cell division in apical meristems and builds *herbaceous stems,* which are nonwoody stems. (For the scoop on apical meristems, see Chapter 3.)

- **Secondary growth increases the width of the stem.** Secondary growth results from cell division of the vascular cambium and builds *woody stems,* stems that contain secondary xylem tissue. (For more info on the vascular cambium and xylem tissue, check out Chapter 3.)

Primary growth

When the cells in the apical meristem divide, they produce primary meristem tissues — the protoderm, ground meristem, and procambium — that will form all the primary tissues of the stem. (For the fundamentals on meristems and tissues, see Chapter 3.)

The organization of primary tissues inside stems differs somewhat among plant groups, but most flowering plants and conifers organize their vascular tissues into *vascular bundles,* groups of pipe-like tissues that run longitudinally through the roots, stems, and leaves.

In primary stems, vascular bundles have three components:

- A layer of thick-walled cells, such as sclerenchyma and collenchyma, that surround the bundle, protecting the bundle and supporting the stem.
- A cluster of phloem cells, located on the side of the bundle closest to the outside of the stem. These cells transport food.
- A cluster of xylem cells, located on the side of the bundle closest to the inside of the stem. These cells transport water and minerals.

Botanists divide flowering plants into two groups — *monocots* and *dicots* — based on several differences in their structures, including differences in stem organization. Monocots include plants such as the grasses, lilies, and palm trees. Dicots include roses, beans, and broadleaf trees. This chapter presents the differences in vegetative structures of monocots and dicots. Chapter 5 covers reproductive differences.

Monocot stems

Most monocots grow only by primary growth, remaining herbaceous throughout their lives. But some monocots, such as palms, have very fibrous leaf bases wrapped around their stems so that they appear to be woody. Figure 4-2 shows the internal anatomy of a monocot stem.

The vascular bundles of monocots form in a spiral arrangement around the stem.

Moving from the outside to the inside, monocot stems, like the one in Figure 4-2, are made up of epidermis and then ground tissue with vascular bundles that appear to be scattered randomly around the stem.

Dicot stems

Many wildflowers and crop vegetables that are dicots grow only by primary growth. Dicots that grow into shrubs and trees begin growing with primary growth and then grow by secondary growth.

Dicot stems, like the one in Figure 4-3, are surrounded by a sheath of epidermal tissue. The rest of the stem, with the exception of the vascular tissue, is made up of ground tissue.

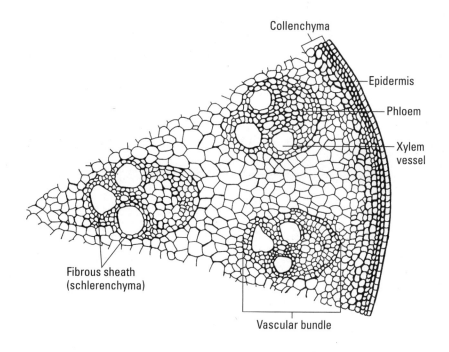
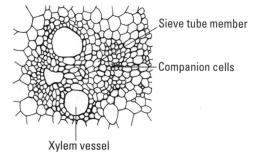

Figure 4-2: Internal anatomy of a monocot stem.

In young dicot stems, ground tissue is organized into two layers of tissue:

- The *cortex* is made of several layers of parenchyma cells and is located between the epidermis and the vascular tissue.
- The *pith* is the group of parenchyma cells in the center of the stem.

Part I: Plant Basics

 Dicots arrange the vascular bundles in a ring around the stem. Moving from the outside to the inside, primary dicot stems, like the one in Figure 4-3, are made up of epidermis, ground tissue (cortex), vascular bundles, and ground tissue (pith).

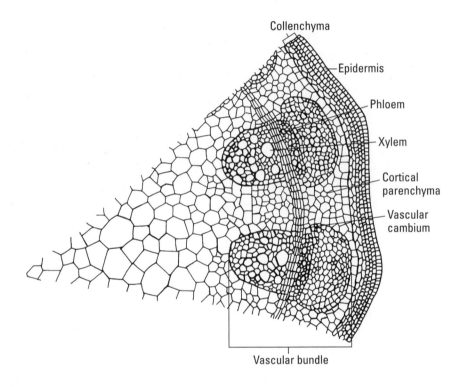

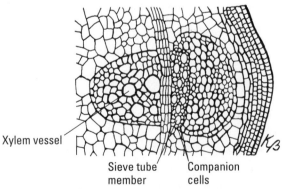

Figure 4-3: Internal anatomy of an herbaceous dicot stem.

Secondary growth

Whether or not a plant does secondary growth depends largely upon how long the plant lives:

- *Annual plants,* which complete their entire life cycle from germination to death within one year, don't do secondary growth.
- *Perennial plants,* which live multiple years, do secondary growth.

Another factor that determines growth pattern is the type of plant. Monocots don't do true secondary growth, while gymnosperms (see Chapter 16) and dicots both do secondary growth in a similar way. Figure 4-4, which shows the secondary growth of a dicot stem, shows an example of how these two groups grow.

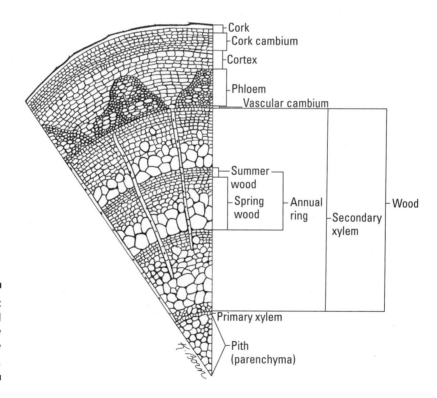

Figure 4-4: Internal anatomy of a woody dicot stem.

Two rings of meristematic tissue, called *lateral cambia,* produce the new cells that make up secondary growth:

- The *vascular cambium* makes new vascular tissue, called *secondary xylem* and *secondary phloem*. As the new cells are made, the vascular cambium pushes the secondary xylem toward the inside of the stem and secondary phloem toward the outside of the stem.

- The *cork cambium* (also called *phellogen)* produces ground tissue (also called *phelloderm)* toward the inside of the stem and *cork cells* (also called *phellem*) toward the outside of the stem. Cork cells help protect woody stems because they're impregnated with a waxy substance called *suberin,* which makes them waterproof and resistant to fire damage, infection, and insect attack. These cork cells eventually replace the epidermal cells that break away as the stem thickens. *Lenticels* are spongy areas within the cork that allow gas exchange.

The bark of a tree consists of all the cells from the vascular cambium to the outside of the stem, so it includes the secondary phloem, parenchyma, and cork cells. The outer bark, which consists of just the cork cambium and the cells it produces (cork and parenchyma), is called the *periderm*.

The vascular cambium and cork cambium both develop from primary tissue in the stem as a stem enters into secondary growth:

- The **vascular cambium** forms from cells between the xylem and phloem within the vascular bundles and from parenchyma cells in the spaces between the bundles. These cells develop into meristematic cells that join up with each other to form a ring of cells just one cell thick within the stem.

- The **cork cambium** usually forms from a ring of cells within the cortex of the stem. These cells develop into meristematic cells that form the ring of cork cambium within the stem.

As the stem grows over time, new cells from the vascular cambium cause the vascular bundles to enlarge, until they eventually merge together to form a central ring of vascular tissue within the stem.

Moving from the outside of the stem to the inside, woody dicot stems, like the one in Figure 4-4, are made up of periderm, including the cork cambium, then ground tissue (cortex), primary phloem, secondary phloem, vascular cambium, secondary xylem, primary xylem, and ground tissue (pith).

As a woody stem grows year after year, the vascular cambium continues to add new vascular tissue. The secondary phloem cells get pushed into the bark of the stem. These cells are fairly fragile and tend to get crushed over time, so the amount of secondary phloem in a woody stem stays relatively constant. However, the secondary xylem is very strong and gets added to the stem in layers, year after year.

The cork is out of the bottle

Wine bottles today sometimes come with screw caps or plastic corks, but the original wine corks are made of — well, cork! Trees called cork oaks, *Quercus suber,* have a well-developed cork cambium and produce large amounts of thick, spongy bark. People harvest this cork sustainably, without harming the trees. Cork forests are the economic livelihood for groups of people in the Mediterranean, and reduced demand for true cork stoppers is hurting these communities.

In temperate climates, the layers of secondary xylem form tree rings:

- ✓ Xylem vessels produced during spring and summer, when growth is most favorable, have a large cell diameter. The wood in this area appears lighter in color and is called *spring wood*.

- ✓ Xylem vessels produced during late summer and fall, when growth is less favorable, have a smaller cell diameter. The wood in this area appears darker in color and is called *summer wood*.

Because each ring represents a year's growth, you can count the rings in a tree to determine how old it is. Botanists and foresters use a tool called an *increment borer* to cut a core of wood out of a tree so that they can count the rings. An increment borer is like a sharp metal straw that can be twisted into the trunk of a tree, cutting the wood as it goes in. When you pull the borer back, it removes a cylinder of wood that shows the tree rings. In addition to telling a tree's age, the core can also tell you which years were good growing years and which years had hazards such as fires. In general, the larger the xylem vessels, the better the year.

Wood is all the tissue inside a woody stem from the vascular cambium to the center off the stem, which is mostly made up of secondary xylem with a small amount of primary xylem. Wood can be divided into two categories, based on whether it's still transporting water and minerals:

- ✓ *Sapwood* has functioning vessels and tracheids. Sapwood makes up the outer part of secondary xylem.

- ✓ *Heartwood* no longer transports water and minerals, but it still provides tremendous support to the stem. Heartwood is impregnated with chemicals called resins and tannins that help prevent infection.

Vascular rays are cells that are arranged horizontally within the wood. They help move materials laterally in stems, from inside to outside.

> ### Senior citizens of the plant world
>
> Small twisted pine trees cling to the rocks on the side of a mountain in the Great Basin National Park in Baker, Nevada, the last of the trees as you climb up to 7,000 feet. It's amazing that these trees can survive the harsh conditions at these high elevations — the cold, the short growing season, and the biting winds. But the trees not only survive, they grow. And just like other woody species, they add a ring of vascular tissue to their stems for every year they're alive.
>
> In 1964, a scientist was given permission to study some of the bristlecone pines growing in the park, including permission to cut down one tree. The scientist cut down a tree, excited to see what he could learn about the history of the tree from looking at its rings. Imagine his astonishment when he counted those rings and found out that the tree was about 4,900 years old! Bristlecone pines that have been studied since the 1960s have confirmed that these trees live to a very ripe old age — the oldest known one that's living today is 4,600 years old — making these the oldest known living trees. And, an interesting insight is that the harsh life at high elevations seems to help bristlecone pines achieve their long life. Bristle cone pines that grow at lower elevations in more pleasant conditions don't live nearly as long.

Specialized stems

You can easily recognize a plant stem when it grows upward, but would you recognize a stem that was growing flat along the ground, or even under the ground?

Some plants modify stems, like the stems shown in Figure 4-5, so that they can serve a special function:

- **Rhizomes** are stems that grow horizontally just at the soil surface or even underground. The rhizome is the main stem of the plant, which periodically produces a new vertical shoots from its axillary buds. Each shoot may look like a brand new plant from above ground, but they're really all identical clones from the original mother plant. If something breaks the rhizome into pieces, each shoot can continue to grow as a new plant, resulting in asexual reproduction of the original plant. (For more information on asexual reproduction, see Chapter 11.) Examples of plants that grow by rhizomes include ginger, asparagus, bamboo, iris, and many ferns.
- **Stolons** are a lot like rhizomes because they're shoots that grow horizontally along the ground surface, produce new plants, and can lead to asexual reproduction. Once difference between stolons and rhizomes is that stolons grow as side shoots off an existing stem, whereas a rhizome is the main stem of the plant. Stolons are often called *runners,* and they're probably most familiar from strawberry plants.

- **Tubers** are the enlarged ends of rhizomes or stolons that are specialized for food storage, usually in the form of starch. A familiar example of a tuber is a potato. Even though people dig potatoes out of the ground, they're not roots — they're actually swollen stems. And the eyes on the potato? Axillary buds!

- **Corms** are the bases of stems that are swollen for food storage. Crocus and gladiolus plants both produce corms. If you buy a bag of crocus or gladiolus "bulbs" to plant in your garden, you're actually buying a bag of corms. (I discuss true bulbs in the upcoming section "Specialized leaves.") Taro, which is used to make poi, is also a corm.

- **Tendrils** are plant structures that are modified to wrap around and cling to things for support. Some tendrils, like the one shown in Figure 4-5, are modified stems, but plants can make tendrils from various plant structures. Evergreen grape is an example of a plant that clings with tendrils it makes by modifying stems.

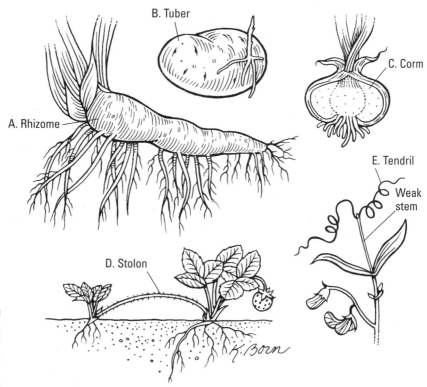

Figure 4-5: Types of specialized stems.

Wood or plastic?

What's safer in the kitchen, a wooden cutting board or a plastic one? Some people think plastic because it's easier to disinfect, but research shows that may not be the case. Studies that compare how long bacteria survive on wood versus plastic cutting boards show that bacteria can actually survive longer on plastic boards, particularly if the cutting boards have knife scars on them. (Scarred plastic boards are hard to disinfect.) The greater safety of wooden boards is no surprise to botanists who know that wood is impregnated with antimicrobial tannins and resins! If these compounds help trees survive for thousands of years by blocking microbial invasion, it makes sense that they'd still be at work in wooden products like cutting boards!

Digging Deep with Roots

Roots anchor a plant into the soil, helping it withstand the push of wind and water. Roots grow through the soil, absorbing the water and minerals plants need and transporting them upward into the shoot through vascular tissue. Plant roots also form very important relationships with soil fungi and bacteria that help plants grow.

When a seed germinates, the embryo produces the first root, called the *radicle*. The radicle grows, producing many branch roots and forming the *root system* of the plant. Root systems for a single plant can be huge in relation to the plant itself. Some plant roots extend deep into the soil; others spread out over a great area.

Figure 4-6 shows the two main types of roots systems:

- **Tap root** systems produce thick main roots that have much smaller branch roots. Some tap roots, like those of a carrot, are swollen because they're specialized for food and water storage. Most dicots make tap root systems.
- **Fibrous root** systems produce many fine roots of similar diameter. Grasses, and most other monocots, have fibrous root systems.

In addition to their main root system, some plants produce additional roots called adventitious roots. *Adventitious roots* are roots produced from a plant organ other than a root. For example, the primary roots of monocots die soon after the seed germinates, and then the plants develop adventitious roots directly from the base of the stem.

Chapter 4: Vegetative Structures: Stems, Roots, and Leaves

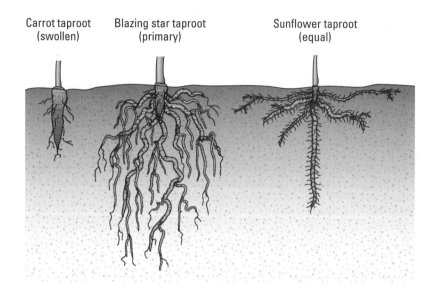

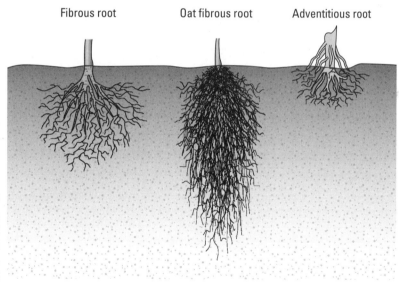

Figure 4-6: Types of root systems.

Regions within roots

Roots grow from their tips, just like shoots, because of an *apical meristem* that produces new cells (see Figure 4-7). The cells of the apical meristem divide frequently, adding new tissue and sending the root telescoping down through the soil. The apical meristem is both valuable and fragile, so plants protect this tissue by covering it with a protective *root cap.* As the apical meristem divides, it pushes cells to both sides, making the root longer and creating the root cap.

A layer of epidermal cells covers the roots, helping protect them and also increasing their ability to absorb water and minerals. Cells of the epidermis can become *root hairs,* single cell projections that extend off the root and increase the surface that's in contact with the soil.

Unlike the epidermis of the stem, the root epidermis has a very thin cuticle so that it can easily absorb water and minerals from the soil. Once water and minerals are absorbed, they're moved into the vascular tissue that forms a *vascular cylinder* in the middle of the root. In most dicots, the primary xylem of the vascular cylinder is arranged in an irregular or cross-shaped pattern. The parenchyma cells of the *cortex* are located in between the epidermis and vascular cylinder. In monocot roots, the vascular bundles form a ring around a central column of pith.

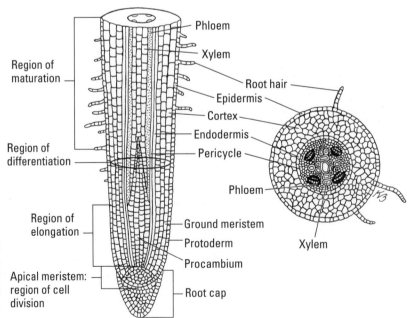

Figure 4-7: The internal anatomy of a dicot root.

Chapter 4: Vegetative Structures: Stems, Roots, and Leaves

Botanists divide roots up into four sections, based on the major activities in each:

- **The root cap is located at the very tip of the root, where it provides protection to the apical meristem.** Cells in the root cap produce slippery molecules that help roots slide through the soil and also supports the growth of bacteria that are beneficial to the plant.

- **The region of cell division is located just behind the root cap.** In this zone, the cells of the apical meristem divide to produce new cells. Just like stems, root apical meristems consist of three types of meristematic tissue:
    - **Protoderm** that produces the epidermis.
    - **Ground meristem** that produces parenchyma cells of the cortex.
    - **Procambium** that produces primary xylem and phloem.

- **The region of elongation is located just behind the apical meristem.** In this zone, the root cells expand and get longer. Small vacuoles inside the cells merge to form the large central vacuoles that are common to plant cells.

- **The region of differentiation** (or maturation) is behind the region of elongation. In this zone, root cells change into their final mature forms. Epidermal cells may become root hairs, for example. If the plant is a perennial, secondary growth may begin in this region.

Controlling transport

Roots have a special layer of cells not found in stems called the *endodermis.* The endodermis is located just at the innermost layer of the cortex and adjacent to the vascular tissue. The endodermis acts kind of like the Great Wall of China — it's a checkpoint for materials entering the plant through the roots.

Endodermal cells have bands of a waxy molecule, called suberin, wrapped around their walls. These waxy bands, called C*asparian strips,* prevent water and dissolved materials from reaching the vascular tissue by moving through the walls of root cells. When solutions spreading through the cell walls of the roots reach the Casparian strip, they have to actually enter into a root cell in order to continue proceed further into the root.

By forcing the solution to enter into the root cell, the plant is able to regulate what types of materials enter. To enter a cell, materials must cross the plasma membrane of the cell. Plasma membranes are very selective and control what is allowed into a cell. (For more on plasma membranes, check out Chapter 2.) Casparian strips also help prevent water loss from the vascular cylinder.

Making lateral roots

The *pericycle* is a layer of cells located at the very outside of the vascular cylinder, just inside the endodermis. The cells of the pericycle, which can consist of one or more layers, remain undifferentiated and capable of cell division even after the cells all around them have matured. The cells of the pericycle divide to produce *lateral roots* that branch off the main root.

It's easy to confuse lateral roots with root hairs, but they are very different. Lateral roots are multicellular structures with all the tissues of a primary root, originating in the pericycle in the middle of the root. Root hairs are single cell projections of the endodermis.

Lateral roots and root hairs expand the network of plant roots that branch through the soil, forming the *rhizosphere,* the area of soil surrounds plant roots. Within the rhizosphere, plants form complex relationships with soil microorganisms that influence how plants obtain water and minerals and can even change the chemistry of the soil itself.

The cells from the outside to the inside of a dicot root are epidermis (root epidermal cells can become root hairs), ground tissue (cortex), endodermis (innermost layer of cortex, contains suberin and casparian strip), pericyle (outer layer of vascular cylinder from which lateral roots originate), phloem (part of vascular cylinder), xylem (inner part of vascular cylinder).

Secondary growth in roots

In woody dicots and conifers, parts of the pericycle and the parenchyma cells around the primary xylem become a vascular cambium. The cells of the root vascular cambium divide to produce secondary phloem to the outside of the root and secondary xylem to the inside. The patches of new secondary vascular tissue merge together with the primary vascular tissue, forming a circular ring of vascular tissue. Each year, the vascular cambium produces new layers of vascular tissue, causing the secondary xylem to form rings in the roots just like it does in the stems.

The pericycle also gives rise to a cork cambium in the roots. The cork cambium produces cork which becomes part of the bark of the root.

Specialized roots

In addition to the primary functions of anchoring plants and absorbing water and minerals, some roots specialize for other functions (see Figure 4-8 for examples):

Chapter 4: Vegetative Structures: Stems, Roots, and Leaves 73

- **Prop roots** are a type of adventitious root that extend from the stem to provide additional support for tall, heavy plants that might have a shallow root system. Corn is a familiar example of a plant that uses prop roots for additional support.

- **Storage roots** store food and water. Beets and carrots, for example, store high amounts of sugar in their roots.

- **Pneumatophores** or breathing roots help plants that grow in very wet areas like swamps get enough oxygen. These roots basically act like snorkel tubes for plant, rising up above the surface of the water so that the plant can get oxygen. Mangrove trees and bald cypress both grow in wet areas and produce pneumatophores.

- **Aerial roots** form when seeds germinate on the branches of other trees, and the growing plant forms roots that wrap around the branch. Aerial roots may grow all the way down until they reach the ground, and sometimes they strangle the host plant. Orchids that grow on other plants produce aerial roots to attach themselves to the plant and to collect moisture and dust from the air. *Buttress roots* are a type of large aerial roots that support tall trees in tropics where they soil is very shallow.

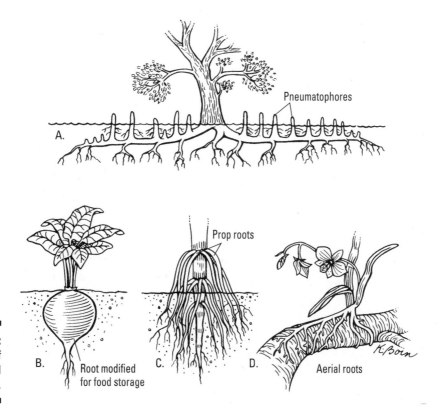

Figure 4-8: Types of specialized roots.

Forming partnerships with soil fungi

Most plants don't work alone to capture water and minerals from the soil. Instead, they form partnerships with fungi called *mycorrhizae*.

The partnership is an example of a *mutualistic symbiosis* that benefits both partners (see Chapter 19):

- **The plants give some sugars from photosynthesis to the fungi.** Basically, the plant is offering the fungi a free lunch, which the fungi are quite happy to accept. The fungi grow all over the surface of the plant roots, and some even penetrate into the plant cells (see Figure 4-9):

    - *Ectomycorrhizae* form a sheath of fungus over the surface of the plant roots, growing between the cells of the cortex to make water and nutrient exchange easier.
    - *Endomycorrhizae*, also called vesicular-arbuscular mycorrhizae, extend little tree-like structures into the cells of the cortex.

- **The fungi help the plants absorb water and minerals from the soil.** The fungi function like an auxiliary root system for the plants, expanding the rhizosphere by absorbing water and minerals and then sharing with the plants.

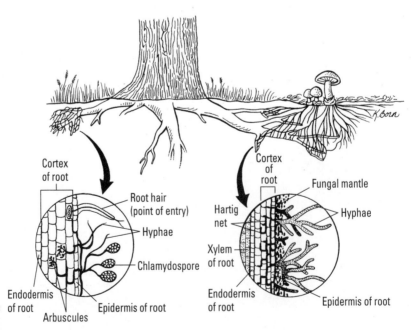

Figure 4-9: Mycorrhizae.

A. Endomycorrnizae (arbuscular)

B. Ectomycorrnizae

Reaching Out with Leaves

Leaves are designed to capture as much sunlight as possible while at the same time protecting the plant's water supply. Most leaves have a broad flat surface that reaches out away from the plant stem in order to intercept the maximum amount of light possible, just like solar panels hanging out there in the sun.

As the sun moves through the sky, plants will even use *solar tracking,* angling their leaves to capture the most sun. When the sun is directly overhead at midday, a bean plant will hold its leaves straight out, turning their flat surfaces directly up toward the sky. Earlier or later in the day, when the sun is at an angle, the plant will let the leaves drop down so that the flat surface is pointed at the right angle to still face the sun.

Leaf structure

Leaves begin as tiny little *leaf primordia,* tiny embryo leaves within the buds of plants. Most leaves have two parts:

- The **blade,** which is the flattened part of the leaf. A network of veins containing vascular tissue runs through the leaf blade.
- The **petiole,** or stalk, of the leaf. Some plants have a pair of appendages, called *stipules,* at the base of the petiole.

Some of the food you eat is made of leaves. When you have a salad, you eat the blades of the lettuce leaves. If you have onions in your salad, you're eating swollen leaves. And if you eat celery, you've eating giant petioles!

Leaves have the same tissues as do stems and roots. Figure 4-10 shows the organization of these tissues inside a leaf:

- **The epidermis is the outer protective layer of cells that cover the leaf.** Leaves need to exchange gas with the environment — for example, they need to take in carbon dioxide for photosynthesis — but they also need to minimize water loss. Two features of leaves help them exchange gases without losing too much water:

    - **A waxy *cuticle* coats the outer surface of the epidermis.** The cuticle helps prevent water loss from the leaf to the air.
    - ***Guard cells* within the epidermis control whether the pores in the leaf, called *stoma,* are open or closed.** So, by controlling the guard cells, plants make sure that they open their stoma only when it's safe and they won't lose too much water. When the guard cells fill up with water, they swell and pull away from each other, opening

the stoma. When the guard cells lose water, they become flaccid and move toward each other, closing the stoma.

- **Parenchyma cells make up most of the interior of the leaf.** These cells form the main photosynthetic region of the plant, called *mesophyll,* which forms the majority of the tissue between the upper and lower surfaces of the leaf. Mesophyll may contain two different types of parenchyma cells:
  - The upper part of the mesophyll consists of tall, columnar cells called *palisade parenchyma.* These cells are packed together directly under the upper epidermis to absorb as much light as possible in their many chloroplasts.
  - The lower part of the mesophyll contains loosely arranged parenchyma cells, called *spongy parenchyma,* that have lots of air spaces between them.

- **Vascular tissue makes up the veins of the leaf.** The vascular tissue brings sugar from the leaves into the stem of the plant and also conducts water from the plant out into the leaves. Each vein contains xylem and phloem tissue surrounded by a layer of protective, thick-walled fiber cells called the *bundle sheath.* The pattern of veins is typically different in monocots and dicots:
  - Monocot leaves usually have *parallel veins,* where the veins line up in parallel lines.
  - Dicot leaves usually have *reticulate,* or *netted, veins,* where the veins form a branching network in the leaf.

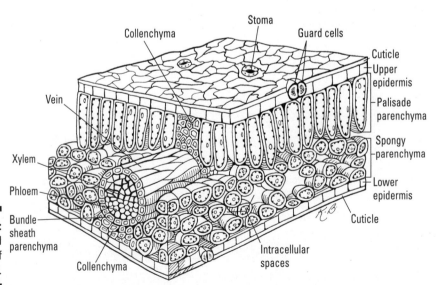

Figure 4-10: Internal anatomy of a leaf.

Leaf types

Leaves come in many different forms and shapes, shown in Figure 4-11:

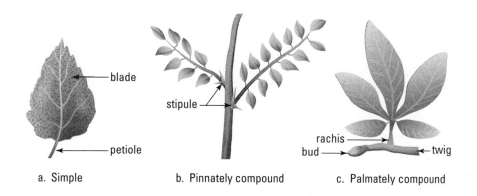

Figure 4-11: Types of leaves and leaf arrangements.

- *Simple leaves* have a single blade.
- *Compound leaves* have a blade that is separated into smaller *leaflets*.

Several different kinds of compound leaves exist:

- *Pinnately compound* leaves have many small leaflets that attach along an axis called a *rachis*.
- *Palmately compound* leaves have several leaflets that attach at the same point so that the leaflets look like fingers on a hand.
- *Double compound* leaves have leaflets that are divided again into more leaflets.

To tell whether a leaf is simple or compound, look for the axillary bud. A simple leaf has a bud in the axil between the leaf and the stem. In a compound leaf, the leaflets might look like individual simple leaves, but they won't have any buds in their axil. The only bud will be found at the base of the entire compound leaf.

Leaf arrangements

Leaves can attach to stems in different ways, forming numerous leaf arrangements. Figure 4-11 shows the three most common *leaf arrangements*:

A cabbage by any other name

Do you like broccoli? How about Brussels sprouts, kale, cabbage, or cauliflower? If you like one, you might like them all — they're all varieties of the same plant. People created each of these vegetables by breeding wild cabbage, *Brassica oleracea,* and selecting for different traits. Just like a farmer might breed his favorite milk cow or a dog fancier selects dogs with a certain trait, farmers chose cabbages that had a unique feature and then bred them to each other over and over again until they developed a variety that looked pretty different from the original wild parent because one feature was greatly exaggerated. When you eat broccoli or cauliflower, you're eating the flowering part of the plant, plus maybe a little stem if eat part of the stalk. When you eat kale or cabbage, you're eating the basal rosette of leaves. And when you eat Brussels sprouts, you're eating giant axillary buds!

- *Alternate leaves* occur when leaves are attached in a staggered arrangement along the stem, with one leaf at each node.
- *Opposite leaves* occur when two leaves attach to the stem directly opposite from each other so that two leaves appear at each node.
- *Whorled leaves* occur when several leaves attach to a stem at a single point so that three or more leaves appear at a single node. When the whorled leaves attach at the base of the plant, botanists call the leaves a *basal rosette*.

Specialized leaves

Some plants modify some leaves either so that they can perform additional functions or to protect the plant from water loss. Figure 4-12 shows some structures that plants form by modifying their leaves:

- *Tendrils* wrap around adjacent stems or other structures. Pea plants are a familiar example of a plant that climbs using tendrils that originate from modified leaves.
- *Reproductive leaves* allow plants to reproduce asexually by producing tiny plantlets along their edges. The plantlets fall off the leaves and grow into new individuals.
- *Bulbs* are clusters of leaves that have fleshy food-storing bases. Onions are an example of edible bulbs.
- *Spines* protect plants from grazers and reduce water loss. The sharp spines on a cactus are the only leaves they produce.
- *Bracts* are leaves that are modified either to protect something or sometimes to attract a pollinator. The showy, red bracts on a *Pointsettia* plant look like petals, but the plant clusters the real flowers in the center of the red bracts.
- *Succulent plants* have thick, fleshy leaves that are modified for water storage. *Aloe vera* is a good example of a succulent plant that's filled with a moist, sticky sap.
- *Cones* and *flowers* also evolved from leaves that were modified for reproduction functions. Petals and cone scales still have a flat leaf-like appearance that suggests their origins.

Figure 4-12: Some structures that plants form by modifying their leaves.

Chapter 5
Reproductive Structures: Spores, Seeds, Cones, Flowers, and Fruits

In This Chapter
- Distinguishing between types of spores
- Discovering the advantages of seeds
- Checking out cones
- Exploring the structure of flowers
- Finding out about fruit

Many plants and other organisms, such as molds, reproduce by simple structures called spores. Two groups of plants, the cone-bearing plants and the flowering plants, evolved an even better reproductive structure — the seed. Flowering plants use flowers to ensure their reproductive success and then package their seeds in fruits that help disperse new plants away from the parent. This chapter presents the structure and function of these plant reproductive structures.

Reproducing with Spores

If you study biology long enough, you'll hear the word spore many times. A *spore* is a reproductive cell that is capable of growing into a new organism or structure without uniting with another cell. Because biologists define the word spore so broadly, lots of different kinds of living things make structures that biologists call spores. You've probably seen black or green mold spores in your kitchen on some old bread, cheese, vegetables, or fruit. Molds, and other types of fungi, reproduce by spores, as do many different kinds of eukaryotic microorganisms. And, as you may have guessed because I'm including them in this chapter, plants make spores as part of their life cycles, too.

Plants make spores during sexual reproduction when plants called sporophytes produce spores by a type of cell division called meiosis. Spores grow into plants called gametophytes. (For more details on meiosis and the life cycles of plants, check out Chapter 11.)

Protecting the Offspring with Seeds

Spores are nice, but seeds are better. (See Table 5-1 for a comparison.) Seeds have protective coverings and food reserves, helping them to survive away from the parent plant. Seeds can remain dormant for hundreds, or even thousands of years, before they begin to grow!

Table 5-1	Plant Spores versus Seeds	
	Spores	*Seeds*
Cells	Single cells	Multicellular containing an embryo, nutritive tissue, and a seed coat
How and where are they produced?	Produced by sporophytes in structures called sporangia by a type of cell division called meiosis.	Seeds develop from ovules, female parts of plants that contain the egg and are fertilized by sperm.
What will it grow into?	A gametophyte	A sporophyte

Seeds *germinate,* or begin to grow, when conditions are right for the new plant. The amount of time seeds remain dormant depends upon a combination of internal signals, such as hormones (see Chapter 10), and external signals from the environment, such the temperature and the availability of light and water. Some seeds germinate within weeks of being released from the parent plant, while others may remain dormant for months to years. Over time, seeds will begin to die if they don't encounter the right conditions for growth.

Two groups of plants evolved the ability to make seeds:

- **Gymnosperms are plants that make seeds, but don't make flowers and fruits**. Gymnosperms often package their seeds in cones. Gymnosperms include pine trees, Ginkgo trees, and cycads. (To explore this plant group, see Chapter 16.)

Chapter 5: Reproductive Structures: Spores, Seeds, Cones, Flowers, and Fruits

- *Angiosperms* **are plants that make seeds, flowers, and fruits.** In other words, angiosperms are the familiar flowering plants (see Chapter 17). Two groups of plants — monocots and dicots — within the angiosperms have distinct structural differences, both in their vegetative and reproductive structures. (For more on these groups, see Chapter 4.) Botanists define these two groups of angiosperms based on the number of seed leaves, or *cotyledons*, that are attached to the plant embryo within the seed:

    - *Monocots* have one cotyledon.
    - *Dicots* have two cotyledons.

Seed structure

Seeds contain new plant embryos along with stored food and are covered by a *seed coat* that protects the seed from destruction. The seeds of each plant group differ in the number of cotyledons and the source of the food for the embryo:

- **Gymnosperms, such as pine, produce embryos with multiple cotyledons that use primary endosperm as food.** *Primary endosperm* develops from maternal tissue that wasn't fertilized so that it contains two sets of chromosomes (is diploid).
- **Dicots produce embryos with two cotyledons.** Some dicots use the cotyledons as food, while most others use a tissue called endosperm. The cotyledons are part of the embryo, so they have two sets of chromosomes (are diploid). *Endosperm* develops from a cell that is fertilized by two sperm, so it contains three sets of chromosomes (is *triploid*). (See Figure 5-1 for a drawing of a dicot seed.)
- **Monocots produce embryos with one cotyledon that use endosperm as food.** (See Figure 5-2 for a drawing of a monocot seed.)

Dicot seeds

If you look closely at a dicot seed, such as a bean or a peanut, you can study its parts (see Figure 5-1):

- The outermost covering is the *seed coat*. If you examine the seed coat carefully, you'll see a couple of tiny marks: the *hilum*, a small scar that marks where the seed was attached to the ovary wall inside the parent plant, and the *micropyle*, a tiny hole that allowed the pollen tube to enter during fertilization.

- If you're looking at a roasted peanut or if you soak a bean seed for a while, you can take the outer covering off and examine the inside part of the seed. The seed will split apart into two halves, which are mostly the two large cotyledons.

- The *embryo* is tucked in between the two cotyledons. If you look closely, you can see a tiny shoot called a *plumule* and the little stub that will develop into the stem and roots, called the *hypocotyl-radicle axis*.

Dicot seeds from peas and beans are a little bit different from other dicots because they belong to the group of dicots that uses the cotyledons as food for the embryo. As a result, the cotyledons in these seeds are very large. Most other dicots use endosperm as food for the embryo. In these seeds, the cotyledons are much smaller and look like tiny leaves at the top of the embryo. The bulk of the tissue surrounding the embryo is endosperm.

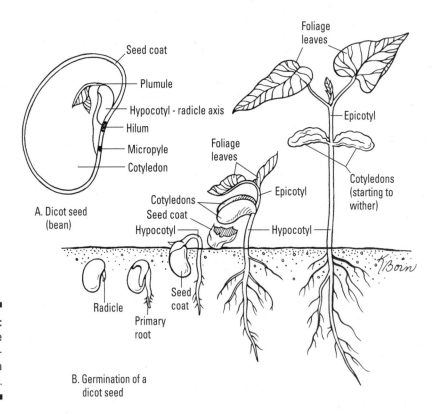

Figure 5-1: Structure and germination of a dicot seed.

Chapter 5: Reproductive Structures: Spores, Seeds, Cones, Flowers, and Fruits   85

As dicot seeds germinate, the food stored in the cotyledons or endosperm provide the growing embryo with energy and building material for growth. Figure 5-1B shows the germination of a bean seed as an example of early dicot development:

1. **The radicle begins to grow and emerges from the seed to form the first root, called the *primary root*.**
2. **The hypocotyl elongates, pushing its way out of the seed and above the soil.**

    As the hypocotyl emerges from the soil, it's bent into a curved shape called the *hypocotyl arch*. The cotyledons protect the upper part of the embryo, called the *epicotyl*, as it emerges from the soil.

3. **Below the soil, secondary roots begin to branch off of the primary root.**
4. **When the hypocotyl emerges into the light, it straightens out and the epicotyl becomes visible, revealing a pair of true leaves and the apical meristem of the plant.**
5. **The leaves expand and begin to do photosynthesis**

    In many dicots, including some beans, the cotyledons turn green and also do photosynthesis until they wither and drop off the plant.

Monocot seeds

To study the parts of a monocot seed, you can take a close look at a corn kernel (shown in Figure 5-2):

- The outermost layer of a corn kernel is the *pericarp*. If you remove the pericarp, the next layer you'll find is the *endosperm* that supplies the embryo with food.
- Like dicots, monocot embryos have a plumule and a radicle. In monocots, however, the plumule is enclosed in a tubular structure called the *coleoptile*, and the radicle is enclosed in the *coleorhiza*. Monocot embryos have only one thin, leaf-like cotyledon.

A corn kernel is actually a type of dry fruit called a *caryopsis*, or *grain*, that contains a single seed. However, the wall of the fruit, called the *pericarp*, is fused to the seed coat, so everything inside the kernel is part of the seed.

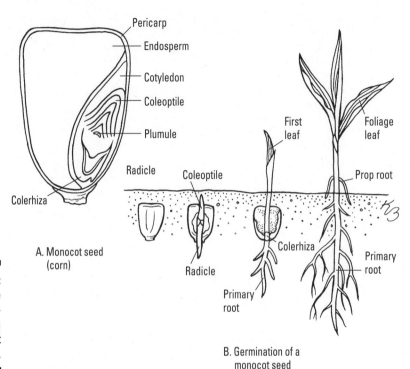

Figure 5-2: Structure and germination of a monocot seed.

The endosperm provides the growing monocot embryo with energy and building material for growth. Figure 5-2B shows the germination of a corn kernel as an example of early monocot development:

1. **The radicle grows first, piercing the kernel and growing downward into the soil to form the primary root.**
2. **The primary leaf of the shoot grows upward, pushing its way out of the kernel and the soil.**

    As the leaf grows, it's protected by the tube-like coleoptile.

3. **Once the primary leaf reaches the light, it pushes through the coleoptile and begins to expand.**
4. **Under the soil, lateral roots begin to develop from the primary root.**
5. **In corn and some other monocots, adventitious roots called prop roots also develop from the stem and push their way into the soil.**

Chapter 5: Reproductive Structures: Spores, Seeds, Cones, Flowers, and Fruits

Organizing Reproduction in Cones

Reproductive structures like cones and flowers evolved from specialized leaves called *sporophylls*. Sporophylls are leaves that produce spore-producing structures called sporangia on their surface. As plants evolved to protect the sporangia and produce spores more efficiently, the leaves with sporangia became more specialized, losing their photosynthetic function and becoming shorter and thicker and more densely packed on the stem. Over time, these tight bunches of sporophylls evolved into the cones and flowers you see today.

The sporophylls of some gymnosperms are spirally arranged into structures called *strobili* or *cones*.

Gymnosperms (see Chapter 16) produce spores that grow into gametophytes right inside the cone. Some cones, such as those of cycads, are bisexual, producing both male and female gametophytes. Other cones, such as those of conifers, are either male or female and produce gametophytes of only one sex. After sexual reproduction occurs, seeds develop inside the cones.

Finding a Mate with Flowers

Plants have sex! You may not have thought about it, but whenever you've enjoyed the beauty of a flower, you've basically been admiring an advertising campaign designed to find a plant a suitable mate. Some plants, like the seedless vascular plants (see Chapter 15) and gymnosperms (see Chapter 16) are more discreet about their sexual reproduction, but the flowering plants really like to dress up and show off.

Flower structure

All flowers have four basic parts derived from modified leaves and a supporting stalk. At the base of the flower, you'll find the stalk of the flower, called the *peduncle*. As the pendunkle produces a flower bud or cluster of buds, it swells to form the base of the flower, which is called the *receptacle*. Other parts of the flower — like the petals — grow in *whorls*, or rings, around the receptacle.

You can see the four basic types of parts of flowers in Figure 5-3:

- **The first, outermost whorl of parts that grows from the receptacle is the *sepals*.** Sepals are petal-like structures that are often green. All together, the whorl of sepals is called the *calyx*. When the flower bud is closed, the calyx helps protect the rest of the flower. Sometimes, sepals have bright colors and look like petals. When the sepals and petals look the same, some botanists call all of them *tepals* instead of sepals or petals.

- **The next whorl of parts is the one you're probably most familiar with — the *petals*.** All the petals together are called the *corolla*. The reason you see so many different looking types of flowers are because of the ways plants modify their corollas to attract different pollinators. (See the section "Moving pollen," later in this chapter.)

Together, the sepals and petals form the *perianth* of the flower.

- **After the petals, flowers have a whorl of male parts called *stamens*.** Flowers can have a few or many stamens. Each stamen has two parts:

    - The little bags of pollen at the tips of the stamens are called *anthers*. Anthers contain sporangia. The spores produced there develop into pollen grains containing sperm cells.

    - The slender stalks that support the anthers are called *filaments*.

- **The most central whorl of parts within a flower consists of female *carpels* — green, vase-like structures.** In many flowers, more than one carpel fuses together to form a compound carpel, also called a *pistil*.

Each carpel or pistil has three parts:

- The tip of the pistil is the *stigma*. Pollination occurs when pollen grains land on the stigma.

- The long, tubular part of the pistil is called the *style*. Pollen grains grow long tubes down through the style in order to reach the base of the pistil.

- The swollen base of the pistil is the *ovary*. Inside the ovary are chambers that contains one or more *ovules*. After the ovules are fertilized with sperm from the pollen, they will develop into seeds. (For more details on fertilization and seed development in flowering plants, check out Chapter 17.) Ovaries with one chamber are *simple ovaries;* ovaries with multiple chambers are *compound ovaries*.

One way that flowers can look different from each other is based on the position of the ovary relative to the calyx and corolla. If the calyx and corolla are attached to the receptacle underneath the ovary, botanists say the ovary is *superior*. Likewise, if the calyx and corolla are attached to the receptacle above the ovary, botanists say the ovary is *inferior*. Figure 5-3 shows a flower with a superior ovary.

The two groups of flowering plants, dicots and monocots, make flowers with different numbers of petals and other flower parts. Dicot flowers typically have parts in fours and fives, or multiples of fours and fives, while monocot flowers typically have parts in threes or multiple of threes.

Botanists call the entire whorl of male parts the *androecium* and the whorl of female parts the *gynoecium*.

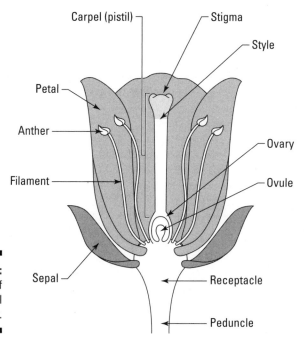

Figure 5-3: Parts of a typical flower.

Moving pollen

The world is full of beautiful and amazing flowers in many shapes and colors. The reason that flowers are so diverse is because each type of flower has a different strategy for spreading and gathering pollen. Some flowers are tailored to a particular animal pollinator, while others use wind to move their pollen around. You can guess the strategy that a flower uses by the way it looks and smells:

- Plants that attract bees to help with pollination often make blue or violet flowers, because those are the colors that bees see best. Some petals even have markings that are invisible to humans but visible to bees.

- Plants that attract moths are often white and strongly scented so that moths can find the flowers in the dark.

- Hummingbirds see red well, so plants that are advertising for hummingbirds often make red flowers and provide sweet nectar as a reward. The petals on some bird-pollinated flowers join together to form a tube-like corolla. The beaks of birds have no problem reaching down into these tubes to get nectar and pollen.

- Some flowers attract beetles or flies as pollinators. The reddish, rotten meat appearance and meat or urine-like smells of these flowers can make them less than popular with people!

- In plants that don't attract animals, such as those that use wind for pollination, the flowers may not have a corolla at all. In grasses, the female parts of the flower have feathery extensions designed to capture pollen as it blows by in the wind.

Flower arrangement

Some plants produce solitary flowers, but others cluster flowers together along a single stem. An *inflorescence* is a cluster of flowers along a single stem or series of stems. The little stems that extend to each flower in an inflorescence are the *pedicels,* and the central stem of an inflorescence is the *rachis.* Rachis is the same term that botanists use for the central stem of a compound leaf (see Chapter 4).

Leaves may grow from the base of the entire inflorescence, or at the base of each flower, or both. Sometimes, the leaves around the inflorescence look different from the rest of the leaves on the plant.

Figure 5-4 shows several different types of inflorescences:

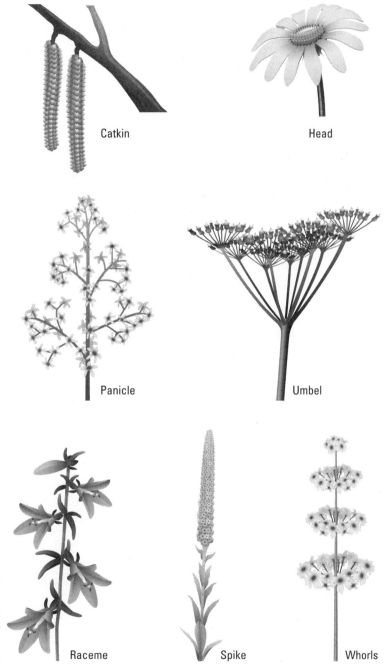

Figure 5-4: Types of inflorescences.

- ✓ **Catkins are drooping cylinders of tightly packed flowers that may lack petals.** Some familiar trees that make catkins are birch, willow, and oak.

- ✓ **A *head*, or *capitulum*, looks like a single flower but is actually a group of many flowers all on a common receptacle.** Some very familiar flowers, such as daisies, asters, and sunflowers, may have been fooling you all these years into thinking you were looking at a single flower, but the part of the flower that looks like petals is actually many tiny *ray flowers*, and the middle of the flower is many tiny *disk flowers*. The disk and ray flowers are arranged on a single receptacle.

- ✓ **Racemes form when a plant produces flowers on short pedicels branching from a main stem or rachis.** The flowers at the base of the main stem bloom first so that the stem is arranged with the oldest flowers at the base and the youngest flowers toward the tip. Wildflowers like foxglove, lupine, and bluebonnets produce their flowers on racemes.

    - *Panicles* **are compound racemes — each axillary stem along the main stem branches into a tiny raceme.** The big, feathery inflorescences of some grasses are panicles, as are familiar garden flowers like lilac and astilbe.

    - *Spikes* **are a type of raceme where the flowers are *sessile*, meaning they sit close to the rachis rather than extending off on pedicels.** Lavender and other mints are examples of plants that produce their flowers on spikes.

- ✓ **Umbels form when a plant makes many short flowering stems that extend from the same point on the main stem, similar to the way the spokes of an umbrella extend out from one point on the shaft.** Like racemes and panicles, umbels can be simple or compound: In a compound umbel, each of the short stems that extend from the main stem branch into miniature umbels called *umbellates*. Queen Anne's lace, or wild carrot, is a familiar wildflower that produces flowers in compound umbels, such as the one in Figure 5-4.

Packaging the Seeds in Fruits

Plants that make flowers also make fruit. Once a flower has been pollinated and the sperm fertilize the eggs in the ovules, the ovary of the flower develops into a fruit. The ovary wall becomes the flesh or outer covering of the fruit, and the ovules develop into seeds.

As the ovary ripens into a fruit, it develops into three separate regions (see Figure 5-5):

Chapter 5: Reproductive Structures: Spores, Seeds, Cones, Flowers, and Fruits

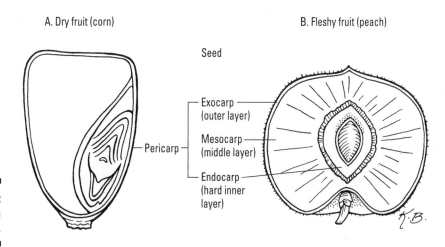

Figure 5-5: Regions of a fruit.

- The skin of the ovary becomes the *exocarp,* or outside of the fruit.
- The innermost layer of the ovary that's closest to the ovules develops into the *endocarp,* or inner boundary around the seeds.
- The *mesocarp* is everything in between the exocarp and the endocarp.

Together, the three layers of a fruit form the *pericarp.*

Fruit types

Fruits come in many different forms from soft and fleshy to hard and dry. And just as some plants produce single flowers and others produce inflorescences, some fruits are made from single ovaries, while others are made from multiple ovaries joined together. Figure 5-6 shows some examples of the many types of fruits produced by flowering plants.

The way most people use the word fruit and the way that botanists use the word fruit can be quite different. In nonscientific terms, people tend to think of fruits as sweet. But to a botanist, any ripened ovary is a fruit. So, while it's true that apples, oranges, and grapes are fruits, so are tomatoes, nuts, cucumbers, string beans, maple "helicopters," grains of grass, and acorns.

Figure 5-6: Types of fruits.

Chapter 5: Reproductive Structures: Spores, Seeds, Cones, Flowers, and Fruits

Fleshy fruits

In fleshy fruits, at least part of the mesocarp is soft or fleshy when the fruit is ripe. *Simple fleshy fruits* form from flowers that have one pistil with either a simple or compound ovary.

- **Drupes are fleshy fruits with a single seed protected by a hard, stony endocarp called a *pit*.** Anytime you've eaten a peach or plum, you've eaten a drupe. Coconuts are an unusual example of a drupe — the husk that many people never see is the exocarp and mesocarp of the fruit. The coconuts sold in stores are just the pit of the coconut fruit with the hard endocarp and seed within.

- **Berries usually develop from compound ovaries and contain multiple seeds.** The endocarp of berries is fleshy like the rest of the pericarp, so it's difficult to tell this layer from the rest. *True berries,* such as tomatoes, grapes, and peppers, have a thin skin and are soft when ripe. *Pepos* are berries with a thick skin, or *rind,* as in pumpkins and other squash. Oranges, lemons, and other citrus fruits are a type of berry called a *hesperidium* that has a leathery, oily skin.

- **Pomes form when the fruit receptacle enlarges and grows up around the ovary, becoming part of the fruit.** The endocarp of pomes forms a papery or leathery layer around the seeds. Apples are probably the most familiar pomes. In apples, the ovary itself becomes the core of the apple, while the rest of the apple develops from the receptacle.

Botanists also call fruits like apples *accessory fruits* because part of the fruit develops from a part of the flower other than the ovary itself.

Aggregate fruits

Aggregate fruits form from single flowers that have multiple pistils. Each pistil develops into a *fruitlet,* and all the fruitlets are clustered together on a single receptacle. Raspberries, blackberries, and strawberries are all aggregate fruits (and not really berries at all!). The part of the strawberry that gets red and juicy is actually swollen receptacle, and the little hard seed-like things on the outside little fruitlets called *achenes*. Because the receptacle forms part of the fruit, botanists call it *accessory tissue*.

Multiple fruits

Multiple fruits develop from multiple flowers in a single inflorescence. The ovaries of the individual flowers develop into fruitlets that grow together to form one, larger fruit. Pineapples are an example of a multiple fruit — each little diamond-shaped section that you see on the outside of a pineapple represents one of the original flowers in the inflorescence. All the fruitlets fused together to form the larger pineapple.

Dry fruits

Dry fruits have a dry mesocarp when the fruit is mature. Some dry fruits, called *dehiscent fruits,* split open when they're ripe, releasing the seeds. Other dry fruits, called *indehiscent fruits,* stay intact.

Botanists group *dehiscent fruits* according to the way they open:

- ✓ **Follicles split open along one side of the fruit.** Milkweed (such as the one in Figure 5-7), columbines, and peonies all produce follicles.

- ✓ **Legumes split open along two sides.** Beans, peas, carob, and mesquite all produce legumes.

- ✓ **Siliques also split open along two sides, but the seeds initially stay attached to a central partition inside the fruit.** Siliques tend to be long and thin. Fruits that open the same way as siliques, but are shorter and wider, are called *silicles*. Silicles and siliques are made by plants in the mustard family, such as broccoli, radishes, and cabbage.

- ✓ **Capsules can split open in a number of different ways.** Some capsules, like those of iris plants, split open along the partitions between the chambers of the ovary. Others form a cap at one end that may pop off to release the seeds. The capsule of poppy flowers, such as the one shown in Figure 5-7, develops holes in the cap through which the seeds can be shaken out.

Botanists group *indehiscent fruits* according to the degree of attachment between the seed and the pericarp and the structure of the pericarp.

- ✓ **Achenes have seeds that are attached to the pericarp only at the base.** Sunflowers produce achenes. Because the seeds are attached to the husk at only one point, it's relatively easy to pry the seeds out and eat them.

- ✓ **Nuts have one seed, but they are larger than achenes, and they have a thick, hard pericarp.** True nuts, such as acorns or filberts, have a cup or cluster of bracts at their base.

    Many foods that people commonly call nuts, aren't really nuts — botanically speaking. Peanuts are legumes; almonds, cashews, pistachios, and macadamia nuts come from drupes; and Brazil nuts are seeds from a capsule.

- ✓ **Grains or *caryopses* have seeds that are tightly joined to the pericarp so that they can't be separated from each other.** All members of the grass family, including wheat, corn, barley, rice, and oats, make grains.

Chapter 5: Reproductive Structures: Spores, Seeds, Cones, Flowers, and Fruits

- **In *samaras*, the pericarp forms wings that extend out from the seeds.** The winged fruits of maple trees are a common example of samaras.
- ***Schizocarps* split into two halves when they are mature, but the seeds remain inside the fruit.** Carrots, dill, and caraway all make schizocarps.

Dispersing Fruits and Seeds

For organisms that can't get up and move around, plants sure seem to get around. Plants have successfully managed to spread themselves over most of the land on planet earth, without ever moving a muscle. Instead, they've evolved many clever strategies for using the resources available to them — wind, water, and animals — to cast their seeds far and wide.

Figure 5-7 summarizes the strategies used by plants to disperse their fruits and seeds:

- **Wind dispersal:** Winged maple seeds spin like helicopters in the breeze, and dandelion fruitlets use plumes like little parachutes to carry themselves far and wide. Plants that use wind dispersal produce light fruits or seeds that may have attachments like wings or hairs to help give them lift.
- **Animal dispersal:** Prickles and burrs and poop, oh my! Some fruits and seeds, such as those of bedstraw and cockleburs, have spines or thorns that catch in animal fur or on hikers' socks, so the plant can hitch a ride to a new location. Other plants make fleshy fruits with seeds that can pass safely through an animal's digestive system. Animals eat the fruit and then spread the seeds around later after digestion.
- **Mechanical dispersal:** Instead of relying upon wind, some plants build explosive mechanisms into their fruits. When witch hazel capsules dry, they split in a way that flings the seeds far from the parent plant. A simple touch will cause the swollen capsules of a type of *Impatiens* called touch-me-nots to explode, helping the plants to disperse their seeds and providing amusement for hikers at the same time.
- **Water dispersal:** Some plants have seeds with waxy coatings or attached sacs of air that help the seeds float so that they can be dispersed by water. Coconuts float in the ocean, dispersing themselves to new tropical islands.

Wind	Animal	Mechanical
Maple	Blackberries	Witch hazel
Dandelion	Acorns	Himalyan balsam
Poppy	Bedstraw	
Butterfly weed		

Figure 5-7: Fruit and seed dispersal.

Part II
The Living Plant: Plant Physiology

In this part . . .

A plant's gotta do what a plant's gotta do, and for plants that means getting energy, building materials, and water. Then they have to move materials around so that the whole plant gets a piece of the action. Plants also have to be able to respond to their changing environment, and grow and develop to complete their life cycles.

In this part, I highlight the fundamental processes plants use to get energy, move materials, and respond to their environment.

Chapter 6

Metabolism: How Living Things Get Energy and Matter

In This Chapter

▶ Getting the big picture on metabolism

▶ Checking out enzymes

▶ Transferring energy and electrons

*J*ust like all living things, plants need energy and matter to fuel their growth and survival. Ultimately, plants get their energy from the sun and the matter they need to build their bodies from carbon dioxide and water. The story of what happens in between the capture of matter and energy from the environment and how a plant cell converts those materials to what the plant needs is the story of metabolism — the sum of all chemical reactions that occur in a cell. This chapter provides a few fundamentals to help you navigate your way through the complicated metabolic processes of photosynthesis and cellular respiration, which the next two chapters cover in detail.

The Big Picture: Overview of Metabolism

You might think of plants as pretty calm organisms, but their cells are just as busy as yours! Plant cells do hundreds of different chemical reactions all at once, such as making hormones, building new structures, creating defensive compounds, and producing pigments for flowers.

Metabolism is all of a cell's chemical reactions added together. Metabolism can be divided into two main categories:

- ✓ **Anabolism includes all the reactions that build larger molecules from smaller ones.** You probably know that plants take in carbon dioxide (CO_2) from the environment. Plants use that CO_2, along with water (H_2O) and energy from the sun, to make sugars. This process, called *photosynthesis,* is an example of anabolism because it makes larger molecules (sugars) from smaller ones (CO_2 and H_2O).

- ✓ **Catabolism includes all the reactions that break larger molecules into smaller ones.** Catabolic reactions allow plant cells to transfer the energy from the sugars they make to a form that's easy to use in the cell.

Plants play a very important role in life on Earth because they can make the food that all living things need to survive.

Food provides all cells, including those of plants, with the *energy* they need to move and grow, as well as the raw materials, or *matter,* they need to build their cells.

Organisms that can make food, like plants, are called *autotrophs* (*auto*=self, *troph*=feed, so plants are literally "self-feeders"). Autotrophs combine matter and energy to make food (see Figure 6-1), and then all organisms — including autotrophs! — use that food to get the matter and energy they need for growth. Organisms that can't make food and have to eat other organisms to survive are called *heterotrophs* (*hetero*=other, *troph*=feed, so heterotrophs like you and me are literally "other-feeders").

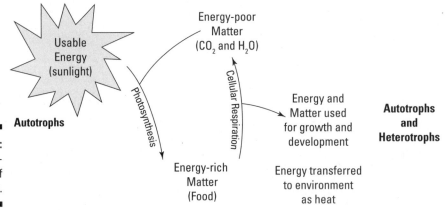

Figure 6-1: An overview of metabolism.

Matter constantly cycles between autotrophs and heterotrophs as a result of their metabolic pathways (see Figure 6-1):

- CO_2 and H_2O are taken in by autotrophs and converted into food and oxygen gas (O_2) through metabolic processes like photosynthesis. (For the details on photosynthesis, flip to Chapter 7.)

- Food is broken down by heterotrophs and autotrophs through metabolic processes like cellular respiration, releasing CO_2 and H_2O back to the environment. (For more on cellular respiration, check out Chapter 8.)

Moving through Metabolic Pathways

Metabolic pathways are series of chemical reactions that convert one molecule into another by rearranging the bonds between the atoms. In cells, chemical reactions usually happen in many small steps rather than one quick change. By doing many small reactions, cells control the energy changes and prevent cellular damage. For example, a cell that needs to make molecule F out of molecule A might do so in five small steps:

$$A \rightarrow B \rightarrow C \rightarrow D \rightarrow E \rightarrow F$$

A represents the starting molecule, or *substrate*. F represents the ending molecule, or *product*. B, C, D, and E all represent molecules that are made during the conversion of A to F; they're called *intermediates*. Every arrow represents one step, or *reaction*, as a chemical change occurs.

One metabolic pathway can connect with another, forming a web of interconnected reactions:

- Some metabolic pathways, called *cyclic pathways*, recreate the initial substrate during the pathway. Making more of the initial substrate allows the cell to repeat the pathway as often as necessary. For example, molecule F is recreated by the cyclic pathway in Figure 6-2.

- The product of one pathway is the substrate of another. For example, in Figure 6-2, product F becomes the substrate for the cyclic pathway.

- A pathway may have one or more branches as intermediates connect with other pathways. For example, in Figure 6-2, molecule H can either get converted into molecule L, or it can be used in the branching pathway and become molecule N.

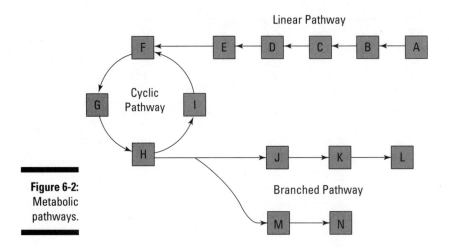

Figure 6-2: Metabolic pathways.

Speeding Things Up with Enzymes

Living things grow, change, and respond at a very fast pace. The chemical reactions that cells need in order to get matter and energy for growth don't happen fast enough on their own to keep up with the pace of life. That's where enzymes come in.

Enzymes are proteins that *catalyze,* or speed up, chemical reactions. You can recognize the name of an enzyme because it usually ends in ASE.

So, when you look at a metabolic pathway, like A→B→C→D→E→F, each arrow represents not just a chemical reaction, but the action of an enzyme that's helping that chemical reaction.

Enzymes have three key characteristics:

- **Enzyme structure is essential to its function.** Enzymes are folded into precise shapes, creating pockets that bind specifically to certain molecules. Molecules that enzymes act on are called *substrates*. The pocket on the enzyme that binds the substrate is called an *active site*. (See Figure 6-3 for an example of an enzyme, its substrate, and the active site.) If an enzyme loses its shape due to environmental changes, it won't work correctly.

- **Enzymes are specific.** Each enzyme can bind only its particular substrate(s). So, when you look at the previous example pathway that converts molecule A to molecule F and you see that five reactions with five different substrates occur, you can figure out that five different enzymes are needed to get the job done. And, if just one enzyme isn't working correctly, the entire pathway will be shut down.

✔ **Enzymes are recycled.** Enzymes assist chemical reactions, but they aren't changed by them. So, the same enzyme can catalyze a reaction over and over again.

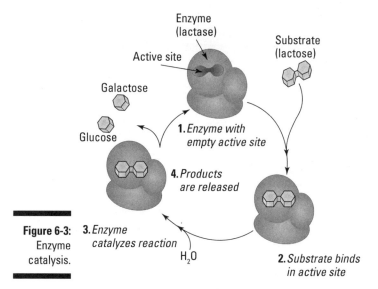

Figure 6-3: Enzyme catalysis.

Transferring Energy with ATP

One fundamental purpose of metabolism is to make energy available to cells. Plants transfer energy from the sun into chemical energy in food and then need to transfer the energy in food to a form that's easy for cells to use.

Cells use the molecule *adenosine triphosphate* (ATP) to store and provide energy for cellular processes like growth, transport, and movement.

ATP gets its name from the three phosphate groups that are attached to the molecule (see Figure 6-4). The negative charges on the phosphate group repel each other, creating a sort of molecular tension, or stored energy, in ATP. When one of the phosphate groups is removed by an enzyme, ATP is converted to *adenosine diphosphate* (ADP). ADP has only two phosphates, resulting in less molecular tension and less stored energy.

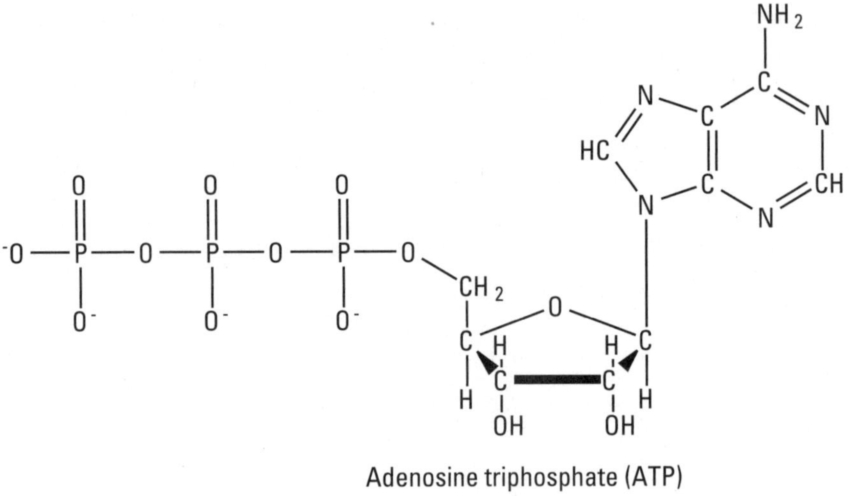

Figure 6-4: The energy carrier adenosine triphosphate (ATP).

Adenosine triphosphate (ATP)

Plant cells constantly make and break ATP, creating an ATP/ADP cycle (see Figure 6-5):

- When plant cells capture energy from the sun or do catabolic reactions that make energy available to the cell, they capture some of that energy by forming ATP molecules.
- When plant cells need energy, they break ATP molecules to ADP, making some energy available to the cell.

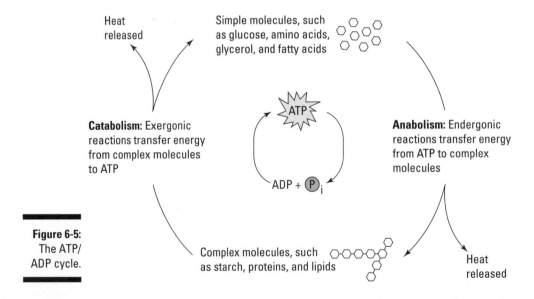

Figure 6-5: The ATP/ADP cycle.

By cycling between ATP and ADP, ATP functions a little like a rechargeable battery for a cell. ADP is like the dead battery. Plant cells can recharge ADP by using energy from the sun or from catabolic reactions. (Check out the left side of Figure 6-5.) When plant cells need energy, they use their ATP batteries, making some energy available to the cell. (See the right side of Figure 6-5.)

Shuttling Electrons with Electron Carriers

Another common event during metabolism occurs when cells use enzymes to transfer electrons from one molecule to another.

Molecules change when they give or take electrons (see Figure 6-6):

- When a molecule gives up an electron, it's *oxidized*.
- When a molecule accepts an electron, it's *reduced*.

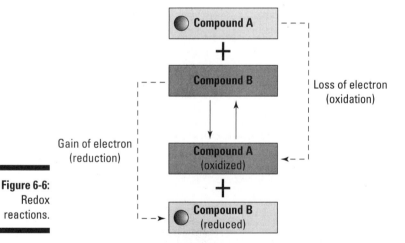

Figure 6-6: Redox reactions.

During metabolism, **oxidation** and reduction reactions occur in pairs called *redox reactions* as electrons are transferred from one molecule to another. The reaction shown in Figure 6-6 is a redox reaction because it combines the oxidation of compound A with the reduction of compound B.

 To help you remember the definitions for oxidation and reduction, try the sentence "LEO the lion goes GER," which stands for Loss of Electrons is Oxidation (LEO); Gain of Electrons is Reduction (GER). It may seem backwards to think that when a molecule *gains* an electron, it's reduced, but if you remember this sentence, you can't go wrong.

During metabolism, cells often use enzymes to transfer electrons from one molecule to a molecule called an *electron carrier* that acts like a shuttlebus for electrons. Electron carriers hold electrons taken during oxidation reactions and then provide them when they're needed for reduction reactions. The electrons are transferred to and from the electron carriers as part of hydrogen atoms (H) that are stripped from one molecule and then given to another.

Just like the energy carrier ATP cycles between its high energy form, ATP, and its low energy form, ADP, electron carriers cycle between two forms (see Figure 6-7):

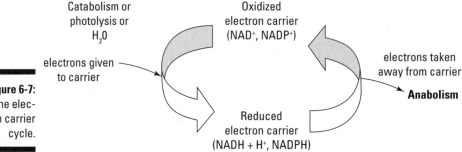

Figure 6-7: The electron carrier cycle.

- ✔ **The oxidized form of a carrier *accepts* electrons from reactions.** For example, plants transfer electrons from water to the electron carrier *nicotinamide adenine dinucleotide phosphate* ($NADP^+$) during photosynthesis. During catabolic reactions such as those that occur as part of cellular respiration, plants transfer electrons to the electron carriers *nicotinamide adenine dinucleotide* (NAD^+) and *flavin adeninine dinucleotide* (FAD). When these electron carriers accept electrons, they convert to their reduced forms.

- ✔ **The reduced form of a carrier *donates* electrons to reactions.** For example, as part of photosynthesis, NADPH donates electrons during the reduction of carbon dioxide to sugar. When electron carriers donate electrons, they convert back to their oxidized forms.

You can easily remember which form of an electron carrier is the oxidized form and which one is the reduced just by looking at its name. When electron carriers are reduced, they're carrying their electron passengers in the form of hydrogen atoms. They show that they are carrying passengers by putting the letter H for hydrogen in their name, as in NADPH. The oxidized form of the electron carrier doesn't have the H, as in $NADP^+$, which reveals that it's not carrying any passengers.

Chapter 7
Photosynthesis: Making Sugar from Scratch

In This Chapter
▶ Discovering the real reason soil is good for plants
▶ Figuring out how to capture energy from the sun
▶ Comparing noncyclic and cyclic photophosphorylation
▶ Exploring the details of carbon fixation

Cells that do photosynthesis have been providing food for life on earth for hundreds of millions of years. During photosynthesis, cells use energy from the sun to make carbohydrates from carbon dioxide. The overall process can be broken down into two major steps, called the light reactions and the light independent reactions. In this chapter, I present an overview of photosynthesis that looks at its importance to life as well as the details of the reactions themselves.

Rethinking the Role of Soil

In order for you to understand photosynthesis, the first thing you need to do is examine what you think you already know about plants and make sure that you're not stuck on a very common misunderstanding.

The most common wrong idea that people have about plants is that they grow by taking matter from the soil. On the surface, this idea seems reasonable. If you look at plants, you see them getting bigger without eating, and their only connection with anything solid appears to be their roots stuck in the ground.

But the power of science lies in taking a closer look at the natural world and carefully testing people's ideas about it. And guess what? When you take a really close look at how plants grow, you find out that it's not really about soil. Yes, plants take water and minerals from the soil, but that's not where they get most of the stuff that makes up the body of the plant. The raw material for building most of the plant body comes from carbon dioxide that plants take in through the stomata in their leaves!

This concept is hard for many people to believe because air seems like nothing, and how could a plant grow from nothing? But air isn't really nothing; it's a mixture of gases. And gases are something. For example, think about steam, which scientists call water vapor. It's water in gas form. If you cool steam, it becomes liquid water, and if you freeze it, it becomes solid ice. Likewise, carbon dioxide, which seems like nothing, can be cooled down to form a solid that people call dry ice. Or, if you're a plant, you can take gaseous carbon dioxide into your leaves and use photosynthesis to rebuild the atoms in the carbon dioxide into carbohydrates it's almost like plants use photosynthesis to make something from "nothing."

van Helmont's experiment

One of the first scientists who investigated the idea that plants take something from the soil was a Belgian doctor named Jean Baptiste van Helmont. In 1642, van Helmont did one of the most famous experiments ever done with plants.

Here are the basics of van Helmont's experiment:

1. **He dried soil in an oven to make sure that it didn't contain any water and then placed 200 pounds of soil in a pot.**
2. **He planted a young willow tree, which weighed 5 pounds, in the pot and moistened the soil with rain water.**
3. **He grew the tree for five years, watering it with rain water or distilled water whenever necessary.**
4. **At the end of the five years, he removed the tree from the pot and weighed it again.**

    At the end of the experiment, the tree weighed 169 pounds, 3 ounces.

5. **He removed the soil from the pot, dried it, and reweighed it.**

    After five years, during which the tree gained 164 pounds, the soil only lost 2 ounces of its weight!

van Helmont's experiment is famous today because it shows clearly what is so hard for people to believe: Plants grow and gain weight while taking hardly anything at all from the soil! At the time, van Helmont thought the mass of the tree must come from water, but now we know that plants are made of carbohydrates, proteins, nucleic acids, and lipids, all of which are carbon-based. So, plants can't possibly grow by water alone; they need a source of carbon which is where carbon dioxide comes into the picture.

The real role of soil

At this point you may be wondering why plants need soil at all. The answer is they don't, really. That's what *hydroponics,* growing plants in a solution of water and minerals, is all about. But notice that hydroponics includes minerals in with the water. That's the key to what soil really provides to plants.

Plants use soil as a source of inorganic compounds called *minerals* and as a matrix in which to grow. Plants get 13 mineral nutrients from the soil. These nutrients can be divided into two categories based on how much plants need:

- *Macronutrients* **are needed in relatively large amounts:**
  - **Nitrogen (N), phosphorous (P)** and **sulfur (S)** are needed for the construction of macromolecules other than carbohydrates. For example, proteins contain large amounts of nitrogen and small amounts of sulfur. If a plant cell needs to build proteins, it takes some of its stored carbohydrates (which contain C, H, and O), adds nitrogen and sulfur from the soil, and rearranges the atoms to form a protein. Similarly, plants can build nucleic acids like DNA, which contain phosphorous and nitrogen, by combining these elements with those from their stored carbohydrates.
  - **Potassium (K), calcium (Ca),** and **magnesium (Mg)** are macronutrients that are needed as helpers called *cofactors* for plant enzymes. Cofactors partner with enzymes to catalyze certain reactions. If enzymes don't have their cofactors, they won't work properly and a metabolic pathway will shut down, causing the plant to grow abnormally.

- *Micronutrients* **are needed in very small amounts.** Micronutrients are sometimes called *trace elements.* Boron (B), copper (Cu), iron (Fe), chloride (Cl), manganese (Mn), molybdenum (Mo) and zinc (Zn) also act as cofactors for plant enzymes.

You can remember the macronutrients needed by plants with the phrase "C. Hopkins café, mighty good!", which is how you read this sentence: CHOPKNS CaFe Mg. C, H, and O stand for carbon, hydrogen, and oxygen, which are the three nonmineral macronutrients needed by plants (and which they get from photosynthesis!). P, K, N, and S are phosphorous, potassium, nitrogen, and sulfur, while Ca stands for calcium, Fe is iron (a micronutrient), and Mg is magnesium.

Soil helps plants grow better because it provides the minerals they need to make molecules that contain elements besides the *nonmineral nutrients*, carbon, hydrogen, and oxygen. Because minerals are also required for optimal enzyme function, you get the best plant growth from soil that is very rich in minerals. To ensure the best growth, people often add chemical fertilizers that contain minerals to the soil. In particular, fertilizers usually contain a great deal of nitrogen and phosphorous, which are the two elements that plants use up most quickly from the soil.

Some fertilizers are called plant food, but they're not really food! They contain minerals like nitrogen and phosphorous that plants tend to use up quickly from the soil. Someone thought "plant food" sounded better than "plant fertilizer," but the only real food that plants use are the carbohydrates they make for themselves.

Discovering Photosynthesis Fundamentals

Photosynthesis transforms light energy into chemical energy that is usable by living things. Plants capture the energy from the sun and store it in the bonds of carbohydrates (such as glucose or G3P, which stands for glyceraldehyde-3-phosphate) they make from carbon dioxide (CO_2) and water (H_2O). With these carbohydrates and some minerals, plants can build all the molecules needed for life: proteins, carbohydrates, lipids, and nucleic acids. Plants also make oxygen gas (O_2) as a waste product during photosynthesis.

Photosynthesis is summarized by the following chemical reaction:

$$6\ CO_2 + 6\ H_2O + \text{light energy} \rightarrow C_6H_{12}O_6 + 6\ O_2$$

This reaction summarizes all the events of photosynthesis into one chemical sentence with reactants (inputs) on the left of the arrow and products (outputs) on the right side of the arrow.

The summary reaction is useful for thinking about the big picture the matter and energy transfers that occur during photosynthesis:

- **Plants transfer matter from the environment into matter inside the plant body.** Plants capture CO_2 from the air and H_2O from the soil and then rearrange the bonds between the carbon, hydrogen, and oxygen atoms, ultimately forming the carbohydrate glucose ($C_6H_{12}O_6$) and the waste O_2. The glucose is used as building material and energy storage inside the plant and the oxygen is released back to the environment.

Chapter 7: Photosynthesis: Making Sugar from Scratch

✔ **Light energy from the sun is transformed into chemical energy and stored in the bonds of sugar for later use by the plant.** Plants use cellular respiration to break down the sugars, transferring the energy to ATP for use in cellular processes. (For the scoop on ATP, see Chapter 6; for more on cellular respiration, see Chapter 8.)

The overall purpose of photosynthesis is to combine matter and energy into food in the form of carbohydrates.

As an analogy, compare a baker making a cake to a plant doing photosynthesis. A baker puts the cake's building blocks, such as eggs, flour, and milk, into a bowl. A plant brings matter in the form of carbon dioxide (CO_2) and water (H_2O) into its cells. The eggs, flour, and milk contain all the necessary ingredients to make the cake, but at this point, they don't look much like a cake. Similarly, the CO_2 and H_2O contain the carbon, hydrogen, and oxygen atoms that are needed to build carbohydrates ($C_6H_{12}O_6$), but they're not yet in the right arrangement. In order to create the changes they want in the cake, the baker places it into the oven. The heat energy from the oven helps the cake batter change. Likewise, plants absorb light energy from the sun and use it to power a series of reactions that build $C_6H_{12}O_6$ from CO_2 and H_2O. Like the heat from the oven, the light from the sun provides the energy to change the raw materials of CO_2 and H_2O into something different — carbohydrates!

Going solar

As people today wrestle with the energy crisis brought on by declining fossil fuels, they're considering alternative energy sources like the sun, an energy source that plants have been relying on for 500 million years. The sun is basically a giant nuclear reactor positioned at a safe distance from the earth nuclear fusion inside the sun produces a mind-boggling amount of energy in the form of *electromagnetic radiation* that includes visible light, X-rays, microwaves, and ultraviolet radiation (see Figure 7-1). Plants capture electromagnetic radiation within the visible light band for photosynthesis.

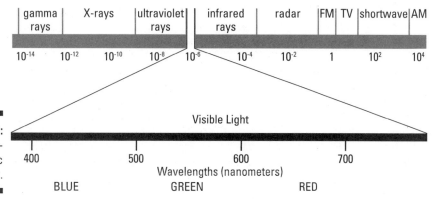

Figure 7-1: The electromagnetic spectrum.

Electromagnetic radiation travels from the sun to the earth in waves. Each type of electromagnetic energy is identified by its *wavelength*, the distance from one energy peak to another. The shorter the wavelength, the more energy in the wave. Gamma rays have the shortest, highest energy wavelengths, while radio waves have the longest, lowest energy wavelengths. Energy waves are measured in nanometers (nm). Each nm is only 1/1,000,000,000 of a meter, so wavelengths of light are pretty small! Plants capture wavelengths between 400 and 700 nanometers for photosynthesis. Radiation with wavelengths of 400 nanometers is visible as blue light, while radiation with wavelengths around 700 nanometers are visible as red light.

Plants primarily absorb red and blue light for photosynthesis

Catching some rays with pigments

Plants appear colored because of their *pigments,* colored molecules that absorb light. The most abundant pigment in plants is *chlorophyll*, which gives plants their green color. Chlorophyll is abundant in plant leaves, where it absorbs light to provide energy for photosynthesis. Chlorophyll appears green because it absorbs blue and red light (see Figure 7-2), but doesn't absorb green light. Green light hits the leaves of the plant and bounces off again, traveling to your eye and causing your brain to interpret the color as green. So, green plants appear green because of their chlorophyll.

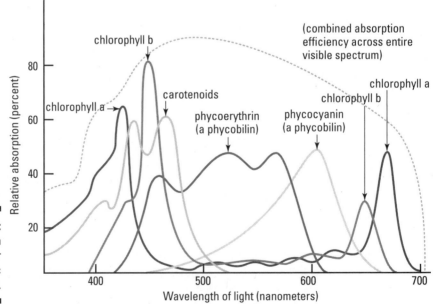

Figure 7-2: Absorption spectra for photosynthetic pigments.

Chapter 7: Photosynthesis: Making Sugar from Scratch

Photosynthetic organisms, such as plants, algae, and some bacteria, contain several types of chlorophyll as well as other pigments. The different types of chlorophyll molecules are all similar in structure, each containing an atom of magnesium (Mg) and long lipid tails that anchor them into the thylakoid membrane of the chloroplast.

The structure of each pigment type determines what kind of light they can absorb (see Figure 7-2):

- Plants, algae, and cyanobacteria (blue-green bacteria) all contain *chlorophyll a*, which absorbs blue and red light and looks blue-green to the human eye.

- Plants and green algae also contain *chlorophyll b*, which absorbs slightly different blue and red wavelengths than chlorophyll a and looks more grass green.

- Brown algae make *chlorophyll c*, which absorbs blue and red and looks green.

- Bacteria other than cyanobacteria produce *bacteriochlorophyll*, which absorbs blue and red and looks very similar to chlorophyll a.

- Plants, algae, and photosynthetic bacteria also contain orange-red pigments called *carotenoids*. Several different pigments, including carotene and fucoxanthin, are in this group. The carotenes in this group absorb blue and look orange. (In which vegetable would you find a lot of these orange-colored pigments? If you said carrots, you're right!) Fucoxanthins absorb green and look yellow-brown. Next time you go to the beach, look for the brownish colored seaweed on the rocks. You'll know that they have fucoxanthin to help them absorb light energy!

- Cyanobacteria and red algae get their colors from *phycobilins*. Phycobilins come in two kinds: *phycocyanin* and *phycoerythrin*. Phycocyanin absorbs in the red and looks blue, giving cyanobacteria their characteristic blue-green color. Phycoerythrin absorbs in the green and looks red. Reddish-colored seaweeds have lots of phycoerythrin.

By having chlorophyll a, plus these other accessory pigments, photosynthetic organisms increase the wavelengths of light energy they can absorb for use in photosynthesis. The dashed line in Figure 7-2 shows all the kinds of light a plant could absorb if it could make all the different types of photosynthetic pigments.

Connections between the light and light independent reactions

Plants use their pigments to capture energy from the sun, ultimately using that energy to transfer electrons from water to carbon dioxide in order to make sugars. Water gives up its electrons as part of hydrogen atoms, leaving behind oxygen atoms that form gaseous oxygen, which is why plants produce the oxygen that you breathe.

During the process of photosynthesis, water is oxidized, and carbon dioxide is reduced (see Figure 7-3).

Figure 7-3: Oxidation and reduction during photosynthesis.

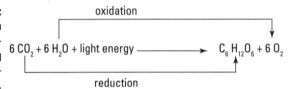

The summary reaction for photosynthesis, shown in Figure 7-3, is a good reminder of the overall purpose of photosynthesis in other words, the "why" of photosynthesis:

- Plants capture matter (carbon dioxide and water) and energy (light) from the environment and combine them to produce food in the form of carbohydrates (glucose).
- As a waste product of photosynthesis, plants release oxygen back to the environment.

Photosynthesis gets more complicated when you tackle the process itself — in other words, the "how." Photosynthesis doesn't occur in one step as indicated by the summary reaction. Instead, it's broken up into lots of little steps which are organized into two separate sets of reactions, called the *light reactions* and the *light independent reactions,* that occur in different parts of the chloroplast. Figure 7-4 presents an overview of the two parts of photosynthesis and shows how they're connected.

The light reactions are also called the light dependent reactions. The light independent reactions are also called the Calvin-Benson cycle, the Calvin cycle, and the Dark Reactions.

Chapter 7: Photosynthesis: Making Sugar from Scratch

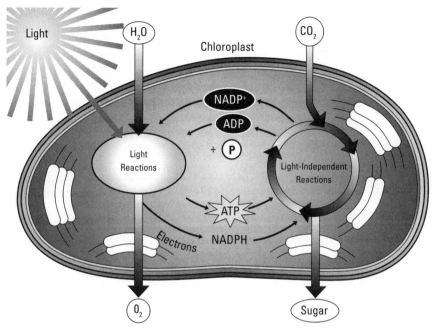

Figure 7-4: The two halves of photosynthesis, the light reactions and the light independent reactions, are separate but linked.

Each part of photosynthesis has a separate purpose, but they both contribute to the overall goal of storing matter and energy in sugar molecules:

- **In the light reactions, plants capture light energy from the sun and transform it into chemical energy.** The energy from light is transformed to chemical energy as it's stored in the energy molecule ATP. During the light reactions, plants also take electrons from H_2O and transfer them to the electron carrier NADPH.

- **In the light independent reactions, plants capture carbon dioxide (CO_2) and convert it into carbohydrates.** In this process, plants use electrons from NADPH and energy from ATP to reduce the CO_2 and convert it to a sugar.

Plants make ATP and NADPH during the light reactions and then use these molecules during the light independent reactions to make carbohydrates from CO_2.

Living things transfer and transform energy all the time. When organisms *transfer* energy, they move it from place to another. If you eat some noodles, you transfer chemical energy from the noodles to chemical energy in your own cells. When organisms *transform* energy, they change it from one type of energy to another. During photosynthesis, plants capture energy from the sun, which is light energy, and transform it into chemical energy in their cells. (Photosynthesis is actually a double whammy — plants transform light energy to chemical energy, and they also transfer the energy from the sun to their own cells!)

To understand the role of electrons and NADPH in the process of photosynthesis, compare the chemical formula for carbon dioxide (CO_2) and the carbohydrate glucose ($C_6H_{12}O_6$). If you want to turn CO_2 into $C_6H_{12}O_6$, what do you need? In other words, what is missing from CO_2? What is missing is hydrogen (H), which is what NADPH provides to the process of sugar construction during the light independent reactions. Essentially, this hydrogen is taken from water (H_2O) during the light reactions and carried by NADPH to the light independent reactions. NADPH donates the H needed to build $C_6H_{12}O_6$ out of CO_2.

Harnessing the Sun: The Light Reactions

The sun releases mind-boggling amounts of energy, absolutely free for the taking, every second of every day. To put the amount of available energy in terms that you can probably relate to, 400 trillion trillion watts of energy reach the earth's surface every second. In one second, that's enough energy to power current human civilizations for 500,000 years. If people tapped into that energy the way that plants do, society wouldn't need fossil fuels at all!

Living cells have been capturing the sun's energy for over 3,000 million years. Bacteria invented photosynthesis, at first using a cyclic set of reactions that didn't produce oxygen. Then cyanobacteria started using water as their electron source about 2,500 million years ago, releasing the first oxygen into the earth's atmosphere. (Imagine that! The very atmosphere of the earth is the way it is today because of photosynthesis!) Eukaryotic photosynthesis performed by the world's first algae probably began about 1,500 million years ago. So, for all these thousands of millions of years, life on earth has relied upon photosynthesis to create the food upon which all living things rely.

Transferring light energy to chemical energy with electron transport chains

Plants absorb light energy and convert it to chemical energy using *electron transport chains*, a series of large proteins stuck in a membrane that pass electrons along the chain like runners in a relay race pass a baton. During photosynthesis, plants use electron transport chains in the thylakoid membranes inside their chloroplasts. When light energy excites the chlorophyll molecules in the chain, electrons start moving from one protein to another. Cells use the movement of electrons through the chain to transfer energy. In the case of the light reactions, energy is transferred from sunlight to ATP. Figure 7-5 shows the electron transport chain plants use during the light reactions of photosynthesis.

Copycat scientists

You might be wondering if photosynthesis is so fabulous and has worked amazingly well for thousands of millions of years, why can't people copy the process to solve energy needs? The good news is that people probably can, and scientists are trying. Photovoltaic cells, the light harvesting part of solar panels, have been in use since the 1950s. Each cell contains a material that absorbs light, causing it to release electrons and create a current. Plants essentially do the same thing during photosynthesis: Chlorophyll absorbs light and then releases electrons to a series of proteins. Most photovoltaic cells in use today use rely on silicon-based materials to absorb the light energy. A couple of problems with silicon-based photovoltaic cells are that they're expensive to manufacture and that they don't perform well on cloudy days.

Recently, scientists in New Zealand have experimented with replacing the silicon-based materials with synthetic dyes, including synthetic chlorophyll! The dyes are cheaper to manufacture than the silicon-based materials and work well in low light similar to that of cloudy days. Other researchers at the Massachusetts Institute of Technology aren't just copying chlorophyll, they're trying to copy the process of the light reactions. The MIT scientists coated viruses with light-harvesting pigments and, then set up a system where the viruses would absorb light energy and interact with catalysts to trigger the splitting of water. The scientists hope to store hydrogen gas from this reaction and develop its use an alternative fuel. Many more groups of scientists around the world are looking to plants for ideas on how to capture sunlight. The future is looking green in more ways than one!

If you think of ATP as a rechargeable battery, then electron transport chains are the battery chargers. During the light reactions, light energizes the battery charger so it can transfer energy into the ATP molecules.

Transferring light energy

Plants maximize their ability to capture light energy by using *accessory pigments* along with chlorophyll. Plants combine these pigments, along with other proteins, to form large complexes called *photosystems* that are part of the electron transport chains in the thylakoid membranes. Each photosystem contains a special molecule of chlorophyll *a* that forms the *reaction center*, the place where light energy is transferred to the electron transport system. Other molecules of chlorophyll *a*, chlorophyll *b*, and carotenoids surround the reaction center chlorophyll. The pigments that surround the reaction center form an *antenna complex* that absorbs light energy and transfers it into these reaction centers.

When the reaction center chlorophyll receives light energy, its electrons are excited to a higher energy state, causing them to move first to outer electron orbitals and then to jump to a protein in the electron transport chain.

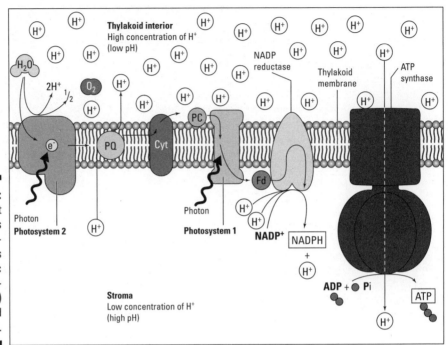

Figure 7-5: The light reactions of photosynthesis (noncyclic photophosphorylation) in a thylakoid membrane.

When energy flows from the antenna complex to the reaction center chlorophyll, it's a little bit like people "doing the wave" at a sporting event. Imagine yourself as the reaction center chlorophyll. Light energy is absorbed by the other pigments, causing them to be energized like a line of people standing up to do the wave. The wave travels nearer and nearer to you and then arrives at your location. You're excited and jump up to do the wave! Lifting your arms is like your electrons moving to a higher orbital. If you were the reaction center chlorophyll, you'd then pass one of these electrons to the electron transport chain.

When the excited electron from chlorophyll is transferred to the electron transport chain, the plant cell has successfully transferred light energy to chemical energy!

Electron transport chains

Electron transport chains accept electrons from an electron donor, pass them along through the protein complexes in the chain, and then pass them to a final electron acceptor.

At the same time the reaction center chlorophyll donates an electron to an electron transport chain in the thylakoid membrane, water molecules are split and electrons from water replace the electrons lost from the reaction center chlorophyll. Water is the electron donor to the electron transport chains during the light reactions.

Electrons move from chlorophyll to the next protein complex in the electron transport chain, which then gives them up to the third complex in the chain. Complexes are temporarily reduced when they accept electrons and then become oxidized again when they give those electrons up to the next member of the chain. So, as the electrons move along the chain, a series of redox reactions occurs. The electrons move from complex to complex along the chain because each member of the chain is more *electronegative,* or has more pull for electrons, than the complex before it. Electrons continue to move through the electron transport chain until they're transferred to the electron carrier NADP$^+$, which is reduced to NADPH + H$^+$.

You can think of the protein complexes in an electron transport chain as a bunch of kids who all want the same piece of candy (the electrons). A little kid is holding the candy. Only a stronger kid (more electronegative) can pull the candy away from the little kid. Then, an even stronger kid can pull the candy away from the second kid, and so on. In this analogy, the kid who ends up with the candy will be the strongest kid of all.

NADP$^+$ is the *final electron acceptor* during the light reactions.

Making ATP by chemiosmosis

Plants use pigments to absorb light energy and then transfer that energy to ATP by a process called *chemiosmosis.* Chemiosmosis uses electron flow through an electron transport chain, coupled with the transport of hydrogen ions (H$^+$) across a membrane to transfer energy to ATP.

The steps of chemiosmosis are

1. **Electrons move from protein to protein in an electron transport chain.**

2. **Some proteins in the chain transfer energy from the movement of electrons to help them pump hydrogen ions (H$^+$) across the membrane.**

3. **The hydrogen ions move back across the membrane through a protein called *ATP synthase,* which can catalyze the formation of ATP.**

4. **ATP synthase transfers energy from the movement of H$^+$ into the formation of ATP.**

To understand chemiosmosis, think about how people use dams to generate hydroelectric power. Water trapped behind the dam is like the H$^+$ trapped behind a membrane. Both represent stored energy. When water crosses the dam through a turbine, the turbine spins, creating electrical current. When H$^+$ flows through ATP synthase, the protein actually spins, too! And just like electrical current powers your home, the flow of H$^+$ through ATP synthase powers the synthesis of ATP.

When plant cells use light energy and electron transport chains to make ATP, it's also called *photophosphorylation*. *Photo-* refers to the light energy that's being used, and *phosphorylation* means adding a phosphate to something In this case, plants cells are adding a phosphate to ADP, which is the building block for ATP. So photophosphorylation literally means "making ATP using light," which is really just another way of saying "the light reactions of photosynthesis."

Noncyclic and cyclic photophosphoryation

Two different types of photophosphorylation occur in plant and other photosynthetic cells:

- **During *noncyclic photophosphorylation,* electrons from water (H_2O) move through an electron transport chain to the electron carrier NADPH.** In other words, the electrons take a one-way journey from water to NADPH. Noncyclic photophosphorylation occurs in plants, algae (seaweed), and cyanobacteria (blue-green bacteria). It's also called the *Z-scheme* because a graph of the energy changes in the electrons as they move through the electron transport chain look like the letter Z (see Figure 7-6). Noncyclic photophosphorylation requires both Photosystem II and Photosystem I. Photosystem II comes before Photosystem I in the Z-scheme.

- **During *cyclic photophosphorylation,* electrons leave chlorophyll when they get excited, travel through an electron transport chain, and return to chlorophyll after their energy has been transferred to ATP.** In other words, the electrons take a round-trip journey from chlorophyll to an electron transport chain and then back to chlorophyll. In plants, algae, and cyanobacteria, cyclic photophosphorylation occurs along with noncyclic photophosphorylation. Bacteria other than cyanobacteria just use cyclic photophosphorylation, which only requires Photosystem I.

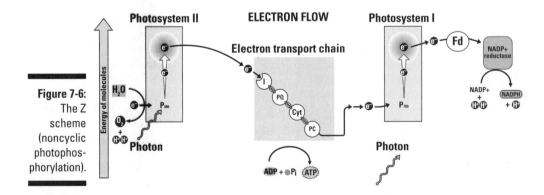

Figure 7-6: The Z scheme (noncyclic photophosphorylation).

Photosystem I and Photosystem II were named for the order in which they were discovered, not for the order in which they appear in the Z scheme. That's why the numbering seems backward in the Z-scheme.

The steps of noncyclic photophosphorylation

During noncyclic photophosphorylation, both photosystems absorb light energy and transfer it to an electron transport chain that's used to make ATP and NADPH. As you read the steps of noncyclic photophosphorylation, refer to Figure 7-6:

1. **Photosystem II absorbs light energy.**

   The accessory pigments transfer light energy to the reaction center chlorophyll, exciting its electrons.

2. **Excited electrons leave the reaction center chlorophyll and move to the electron transport chain.**

   The electrons first pass to *pheophytin,* a molecule similar to chlorophyll, and then to plastoquinone (PQ). Plastoquinone transfers the electrons to an electron transport chain that contains *cytochromes* (cyt), iron-containing proteins that are good at transporting electrons.

3. **The light energy absorbed by Photosystem II also splits a water molecule in a process called photolysis.**

   Electrons from water replace the electrons lost by chlorophyll in Photosystem II. Plants release the oxygen left over from the split water molecules to the atmosphere as oxygen gas.

4. **Proteins in the electron transport chain pump protons across the thylakoid membrane into the interior of the thylakoid.**

   As electrons travel through the chain, plastoquinone (PQ) actively moves H$^+$ across the membrane.

5. **Protons flow back out of the thylakoids through ATP synthase.**

   ATP synthase transfers energy from the movement of H$^+$ to the chemical energy of ATP as it brings together ADP and P.

6. **Electrons move from the cytochromes to the reaction center of Photosystem I.**

   From the cytochrome chain, the electrons move to plastocyanin (PC) and then to photosystem I.

7. **Photosystem I absorbs light energy and transfers energy to the reaction center chlorophyll.**

   The electrons get another boost of energy and leave the reaction center chlorophyll.

8. **The enzyme *NADP reductase* transfers the electrons to the electron carrier, NADP+, reducing it to NADPH + H+.**

   Electrons move from photosystem I to ferredoxin (Fd) then to NADP reductase. Once the electrons move to NADPH, noncyclic photophosphorylation is complete.

Oxygen (O_2) produced by photosynthesis doesn't come from carbon dioxide (CO_2). Oxygen is produced when water is split to provide electrons to the light reactions. Thus, the O_2 comes from the splitting of two H_2O molecules.

The steps of cyclic photophosphorylation

Photosystem I works with Photosystem II to form the Z scheme, but it can also work independently to produce ATP. When Photosystem I works on its own, it does cyclic photophosphorylation.

Here are the steps of cyclic photophosphorylation (follow along in Figure 7-7):

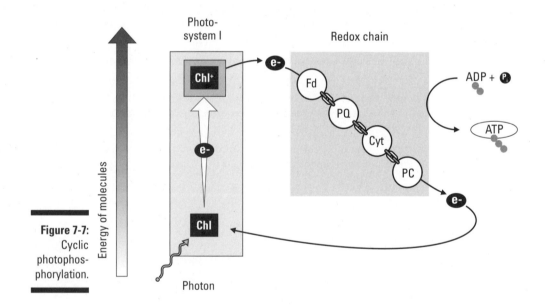

Figure 7-7: Cyclic photophosphorylation.

1. **Photosystem I absorbs light energy.**

   The accessory pigments transfer light energy to the reaction center chlorophyll, exciting its electrons.

2. **Excited electrons leave the reaction center chlorophyll and move to the electron transport chain.**

   The electrons first move to ferredoxin (Fd) just like they would during noncyclic photophosphorylation. But when the electrons leave Fd, they move to PQ (not to NADP reductase). From PQ, they move to the cytochrome chain, and then to PC.

Chapter 7: Photosynthesis: Making Sugar from Scratch

 3. **Proteins in the electron transport chain pump protons across the thylakoid membrane into the interior of the thylakoid.**

   As electrons travel through the chain, plastoquinone (PQ) actively moves H^+ across the membrane.

 4. **Protons flow back out of the thylakoids through ATP synthase.**

   ATP synthase transfers energy from the movement of H^+ to the chemical energy of ATP as it brings together ADP and P.

 5. **Electrons return to their original reaction center.**

   The electrons return to chlorophyll, and the cycle can begin again.

Cyclic photophosphorylation produces ATP but no NADPH + H^+.

Storing Matter and Energy in Sugar: The Light Independent Reactions

For living things to grow, repair themselves, and reproduce, they need a source of building materials. The ultimate source of building materials for most life on earth traces back to one process: the light independent reactions. The *light independent reactions* transform inorganic carbon in the form of carbon dioxide (CO_2) into organic carbon in the form of carbohydrates. All living things build their cells using organic carbon that ultimately traces back to the light independent reactions or similar processes.

The light independent reactions don't directly require light, but they do require the ATP and NADPH made during the light reactions. So, without light, the light independent reactions will eventually shut down.

The steps of the light independent reactions

The light independent reactions occur in the stroma of the chloroplast inside of plant cells.

Like most metabolic pathways, the light independent reactions contain many small steps that lead to the production of carbohydrates from carbon dioxide. Scientists organize the light independent reactions into four major events, shown in Figure 7-8:

 1. **Carbon fixation:** The enzyme rubisco attaches carbon dioxide molecules to a 5-carbon sugar called ribulose bisphosphate (RuBP).

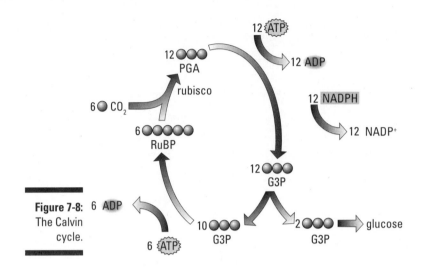

Figure 7-8: The Calvin cycle.

This critical first step captures the carbon dioxide from the environment and is called *carbon fixation*. The 6-carbon molecules that are formed by this step immediately split into two copies of a 3-carbon molecule. These 3-carbon molecules are called *3-phosphoglycerate* (PGA).

Rubisco is a much-needed nickname for the enzyme called ribulose-1,5-bisphosphate carboxylase/oxygenase.

2. **Reduction: Plants use energy from ATP and electrons from NADPH to create carbohydrates.**

   A phosphate group from ATP is transferred to PGA. Next, electrons from NADPH are transferred to the molecule, reducing it to the sugar *glyceraldedehyde-3-phosphate* (G3P).

3. **Biosynthesis: Some G3P is used to make glucose and other sugars.**

   Plants can combine 2 G3P molecules to make one molecule of glucose, then string glucose molecules together to form larger carbohydrates like starch and cellulose. Starch can be stored as a long term source of matter and energy.

4. **Regeneration: Some G3P and ATP are used to recreate RuBP.**

   Plants rearrange bonds between atoms in G3P to rebuild the 5-carbon sugar needed to restart the cycle.

Getting carbon dioxide to rubisco

Getting CO_2 to rubisco, the enzyme that captures carbon dioxide during photosynthesis, can be a tricky business for a plant. The way into the plant for CO_2 through the stomata is also the way out for H_2O. So, on hot sunny days, keeping your stomata open can be dangerous. So, what can a plant that lives in a hot sunny climate do? Close its stomata so that it doesn't dry out, but risk starvation because it can't make food?

And, it gets worse! If a plant does close its stomata to prevent water loss, the concentration of CO_2 in the leaf goes down as the plant uses it up. When CO_2 concentrations get low, rubisco starts to make mistakes. Instead of grabbing CO_2, it will capture O_2 instead, combing O_2 with RuBP in a process called *photorespiration*. Photorespiration produces a toxic molecule that the plant must neutralize, costing the plant energy in the process.

Two groups of plants, called *CAM plants* and *C_4 plants,* have worked out different solutions to the problem of getting carbon dioxide to Rubisco without losing too much water and without setting up conditions that favor photorespiration.

Both CAM and C_4 plants initially capture carbon via carbon fixation using the enzyme PEP carboxylase, which has a stronger affinity for CO_2 than does rubisco. So, in conditions that would normally cause rubisco to be inefficient, these plants can still capture the CO_2 they need.

Crassulacean acid metabolism (CAM) photosynthesis

Many plants that live in desert areas with high-light intensity and dry climate, including cacti and other succulents, protect their water supplies by opening their stomata at night when it's cooler. The CO_2 enters their cells, interacts with PEP carboxylase, and is attached to organic acids for temporary storage during the night. In the daytime, when light is available for photosynthesis, the acids release the CO_2 to Rubisco.

Figure 7-9 shows the steps of CAM photosynthesis:

1. **Stomata open at night to allow CO_2 to enter the plant and combine with phosphoenolpyruvate (PEP), producing 4-carbon acids.**

   The enzyme PEP carboxylase catalyzes the reaction between CO_2 and PEP. This reaction produces malic acid and oxaloacetic acid.

2. **The plant stores the 4-carbon acids in the vacuole of the cell.**

   The acids remain in the vacuole during the night.

3. **During the day, the 4-carbon acids break down, releasing CO_2 into the cell.**

   The plant captures the CO_2 and converts it into carbohydrates using the light independent reactions.

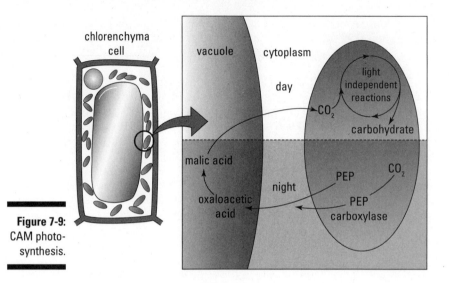

Figure 7-9: CAM photosynthesis.

C_4 photosynthesis

Many tropical plants use a system called C_4 photosynthesis to ensure high concentrations of CO_2 inside their cells even when stomata are closed on a hot day. These plants capture CO_2 in cells near the surface of the leaf where CO_2 is most available and then transport the CO_2 to cells in the interior of the leaf, concentrating it for use in the light independent reactions.

Figure 7-10 shows the steps of C_4 photosynthesis:

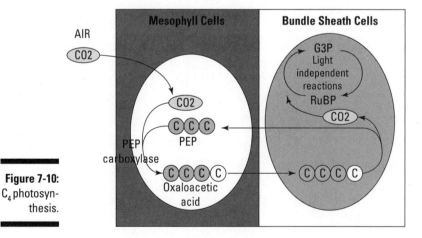

Figure 7-10: C_4 photosynthesis.

1. **Mesophyll cells in the exterior of the leaf attach CO_2 to a PEP molecule, producing 4-carbon acids.**

   The enzyme PEP carboxylase catalyzes the reaction between CO_2 and PEP, producing oxaloacetic acid.

2. **The 4-carbon acid travels through plasmodesmata to bundle sheath cells near the interior of the leaf.**

   The bundle sheath cells produce all the enzymes necessary to do the light independent reactions.

3. **The 4-carbon acid breaks down, releasing CO_2 into the bundle sheath cells.**

   C_4 plants create a high concentration of CO_2 in the bundle sheath cells, ensuring maximum success for rubisco.

Plants that fix carbon directly with rubisco, producing 3-carbon molecules are called C_3 plants because the first molecules they make after carbon fixation are 3-carbon molecules. Plants that fix carbon using PEP carboxylase to produce 4-carbon acids that are transported to bundle sheath cells are called C4 plants because the first molecules they make after carbon fixation are 4-carbon molecules. Plants that fix carbon using PEP carboxylase and then store acids in the vacuoles overnight are called CAM plants, which stands for Crassulacean acid metabolism, after the plant family, the Crassulaceae, in which this pathway was first discovered.

Chapter 8

Cellular Respiration: Making Your Cake and Eating It, Too

In This Chapter

▶ Realizing that plants do cellular respiration, too

▶ Understanding the purpose of cellular respiration

▶ Discovering the three stages of cellular respiration

▶ Digging into the details of glycolysis, the Krebs cycle, and oxidative phosphorylation

*P*lants capture light, carbon dioxide, and water from the environment, using photosynthesis to combine the matter and energy into food molecules, such as carbohydrates. Carbohydrates, such as starch, are a convenient way for plants to store matter and energy for future use. When plants want to access their stored matter and energy, they need to break down food molecules. To break down food, plants use the same process as people — cellular respiration! In this chapter, I present an overview of cellular respiration as well as the details of each stage of the process.

Digging into Cellular Respiration Fundamentals

Plants are famous for their ability to do photosynthesis (see Chapter 7), but very few people know they do cellular respiration, too. But if you stop to think about it for a minute, you'll probably see why it makes sense that plants use both processes:

   ✔ Photosynthesis captures energy from the sun and matter in the form of carbon dioxide (CO_2) and water (H_2O), storing the energy and matter in food molecules, such as the carbohydrate glucose ($C_6H_{12}O_6$).

✔ Cellular respiration breaks down food molecules, allowing cells to access the stored energy and matter so that they can use them to grow, repair themselves, and move — all the things plants need to do to sustain life. When food molecules are completely broken down by cellular respiration, the atoms from the food molecules are released as CO_2 and H_2O.

Cellular respiration and photosynthesis are like yin and yang to a plant — they complement each other. Another way of seeing this relationship is by looking at the summary reaction for cellular respiration and noticing how it's the opposite of the summary reaction for photosynthesis (see Chapter 7):

$$C_6H_{12}O_6 + 6\ O_2 \rightarrow 6\ CO_2 + 6\ H_2O + energy$$

You can think of photosynthesis as the way plants make lunch — storing away some food for when they need it. Cellular respiration is how plants eat that lunch — taking it out, breaking it down, and getting energy and building blocks from stored carbohydrates.

Cellular respiration is a little bit like burning a marshmallow. Food molecules are broken down into CO_2 and H_2O, and energy is released to the cell. When you burn a marshmallow, the food breaks down so quickly you can see and feel the energy as the light and heat from the flames. If cells broke down food molecules that quickly, they'd be cooked! So, cells have to control the burn, breaking down food molecules slowly in many small steps, releasing the energy gradually. Figure 8-1 compares the energy released from the direct burning of sugar, such as the sugar in a marshmallow, with the energy release from cellular respiration.

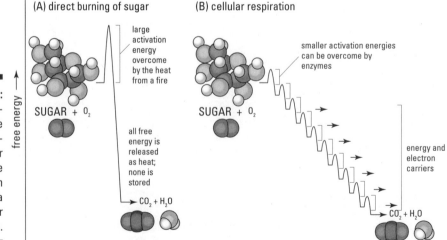

Figure 8-1: A comparison of the direct burning of sugar with the breakdown of sugar via cellular respiration.

Chapter 8: Cellular Respiration: Making Your Cake and Eating It, Too

Cellular respiration is a complicated metabolic pathway, but it's easier to understand if you remember the basic principles of metabolism (see Chapter 6). Like all metabolic pathways, cellular respiration includes many chemical reactions, including redox reactions that move electrons from one molecule to another, and reactions that transfer energy.

Here are some highlights of cellular respiration:

- During cellular respiration, electrons are removed from food molecules, transferred to the electron carrier NADH and then to oxygen (O_2).
- By the end of cellular respiration, the energy from food has been transferred to the energy carrier, ATP.
- Through many chemical reactions, cells completely rearrange food molecules into the waste products, carbon dioxide (CO_2), and water (H_2O).

During cellular respiration, food molecules, such as glucose ($C_6H_{12}O_6$), are oxidized, and oxygen (O_2) is reduced (see Figure 8-2).

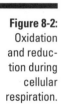

Figure 8-2: Oxidation and reduction during cellular respiration.

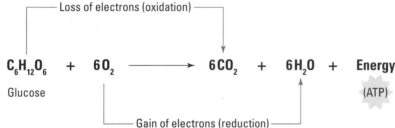

Figure 8-2 shows the overall picture of oxidation and reduction during cellular respiration: Electrons from glucose eventually get handed over to oxygen. However, the actual metabolic pathway occurs in many small reactions that can be divided into three stages (see Figure 8-3):

- **Glycolysis:** Cellular respiration in a eukaryotic cell begins with glycolysis, which literally means the breakdown of glucose (*lyse* means to break). Enzymes in the cytoplasm of the plant cell rearrange the atoms in glucose, forming two molecules of pyruvate. Enzymes in glycolysis also transfer energy from glucose to ATP and electrons from glucose to electron carriers.

✓ **The Krebs or citric acid cycle:** The pyruvate molecules from glycolysis move into the mitochondrion of the plant cell, where enzymes convert them into a molecule called acetyl-coA. Acetyl-coA enters a cyclic pathway that slowly breaks it down, releasing CO_2 as waste. During the Krebs cycle, enzymes transfer some energy to ATP and lots of electrons to electron carriers.

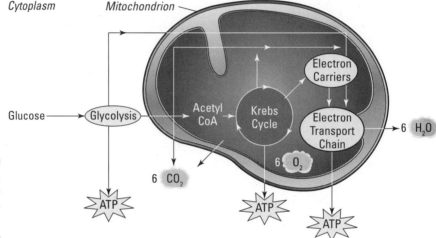

Figure 8-3: An overview of cellular respiration.

✓ **Oxidative phosphorylation:** The reduced electron carriers from glycolysis and the Krebs cycle travel to the inner membrane of the mitochondrion, which contains many electron transport chains. The carriers deliver electrons to the chains, providing the energy for the formation of ATP through chemiosmosis (see Chapter 7).

Breaking Down Glucose in Glycolysis

Glycolysis is a series of ten chemical reactions that oxidize glucose, transfer energy to ATP, and transfer electrons to the electron carrier NAD^+.

The most important events in glycolysis are as follows:

✓ **Enzymes break the bonds between the atoms in glucose, forming intermediates and then creating two molecules of pyruvate.** A glucose molecule has six carbon atoms, whereas a molecule of pyruvate has three carbon atoms, so in a way, glycolysis breaks glucose molecules in half.

✓ **Enzymes transfer energy from intermediates to ATP.** During the first part of glycolysis, enzymes actually break down ATP, transferring energy to the intermediates. But then, in the second part of glycolysis, enzymes transfer even more energy from the intermediates to ATP, leading to a net gain of ATP molecules. The first part of glycolysis, called the *energy investment phase*, uses two ATP but the second part of glycolysis, called the *energy payoff phase,* makes for ATP. Because 4-2=2, glycolysis produces a net gain of two ATP for the cell.

 ✓ **Enzymes oxidize glucose and reduce NAD⁺.** Enzymes transfer electrons that were originally in glucose from an intermediate to NAD⁺, reducing the electron carrier to NADH + H⁺. (See Chapter 6 for the details on NAD⁺.)

For every glucose molecule broken down in glycolysis, the reactions produce 2 ATP, 2 NADH +2H⁺, and 2 molecules of pyruvate.

Glycolysis is universal

Every cell on earth, including plants, animals, and bacteria, needs to break down food molecules to get energy and matter for growth. Even plants, which make their own food, need to break it down to access their stored matter and energy. So, all cells have the ability to do some form of glycolysis. (In fact some cells, such as bacteria and yeast, can get what they need to live from just glycolysis in a process called fermentation.) Variations in glycolysis exist among prokaryotes, but it seems that glycolysis is a necessary part of life on earth. And the fact that glycolysis is so widespread among many different types of cells suggests that it developed very early in the history of life on earth.

Making ATP by substrate-level phosphorylation

When cells make ATP during glycolysis, they use a process called *substrate-level phosphorylation* (see Figure 8-4).

During *substrate-level phosphorylation*, cells make ATP by transferring a phosphate group from an intermediate in a metabolic pathway to ADP. An enzyme binds the intermediate and ADP in its active site and then catalyzes the transfer of the phosphate from the intermediate to ADP.

To remember how substrate level phosphorylation works, think about the words: *Substrate* refers to molecules that enzymes bind in their active sites, and *phosphorylation* means to add a phosphate to something — in this case, adding a phosphate to ADP to make ATP.

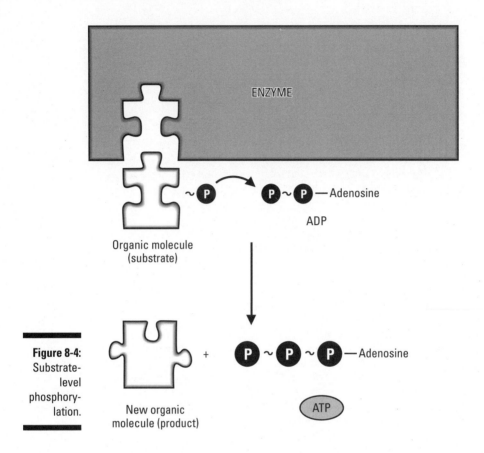

Figure 8-4: Substrate-level phosphorylation.

The steps of glycolysis

Figure 8-5 shows the ten chemical reactions of glycolysis:

1. **An enzyme transfers a phosphate from ATP to glucose, creating the intermediate glucose-6-phosphate and releasing ADP as waste.**

    This step transfers energy to the intermediate and is part of the energy investment phase of glycolysis.

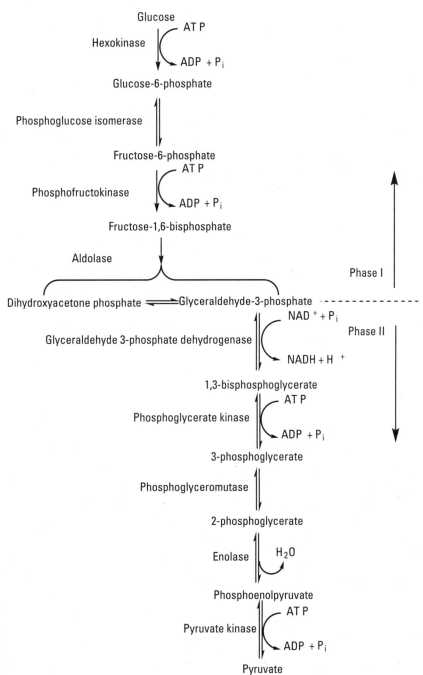

Figure 8-5: The steps of glycolysis.

2. An enzyme rearranges the bonds in glucose-6-phosphate, creating the intermediate fructose-6-phosphate.

3. An enzyme transfers a phosphate from ATP to fructose-6-phosphate, creating the intermediate fructose-1,6-bisphosphate and releasing ADP as waste.

   This step transfers energy to the intermediate and is part of the energy investment phase of glycolysis.

4. An enzyme catalyzes the splitting of fructose-1,6-bisphosphate into two 3-carbon molecules — dihydroxyacetone phosphate and glyceraldehydes-3-phosphate.

   Both of these intermediates ultimately proceed through the rest of glycolysis, so every step after this one happens twice for every molecule of glucose.

5. An enzyme rearranges the bonds in dihydroxyacetone phosphate, converting it to glyceraldehyde-3-phosphate, so that it can proceed through the rest of glycolysis.

6. Electrons are transferred from glyceraldehyde-3-phosphate to NAD^+, resulting in the formation of $NADH + H^+$; at the same time, an inorganic phosphate group that was available in the cell is transferred to glyceraldehydes-3-phosphate, resulting in the formation of the intermediate 1,3-biphosphoglycerate.

7. An enzyme transfers a phosphate group from 1,3-biphosphoglycerate to ADP, producing ATP and a new intermediate, 3-phosphoglycerate.

   This step represents a transfer of energy from 1,3-biphosphoglycerate to the energy carrier, ATP.

8. An enzyme rearranges the bonds in 3-phosphoglycerate, creating 2-phosphoglycerate.

9. An enzyme catalyzes the removal of water from 2-phosphoglycerate, forming the intermediate phosphoenolpyruvate.

10. Another energy transfer occurs as a phosphate group is transferred from phosphoenolpyruvate to ADP, producing ATP.

    This step represents a transfer of energy from phosphoenolpyruvate to the energy carrier, ATP. During the process, phosphoenolpyruvate is converted to pyruvate.

Substrate-level phosphorylation occurs in Steps 7 and 10 of glycolysis.

Going Farther with the Krebs Cycle

The Krebs cycle, or Citric Acid Cycle, picks up where glycolysis left off and involves many of the same types of reactions. Pyruvate made in glycolysis travels into the mitochondrion where enzymes convert it into an intermediate that enters the Krebs cycle. (For a refresher on intermediates and metabolic pathways, see Chapter 6.)

During the Krebs cycle, enzymes oxidize intermediates that came from food molecules, transferring electrons to electron carriers. Some energy is also transferred to ATP.

The most important events in the Krebs cycle are

- **Enzymes oxidize intermediates and reduce electron carriers.** Oxidation of intermediates is the real work of the Krebs cycle. Multiple redox reactions transfer electrons from intermediates in the cycle to the electron carriers NAD+ and FAD. When FAD accepts electrons, it's reduced to $FADH_2$.

- **Enzymes break bonds and rearrange atoms in the intermediates, removing carbon and oxygen atoms and releasing them as CO_2.** For every molecule of pyruvate that leaves glycolysis and breaks down in the Krebs cycle, cells release three CO_2 molecules to the atmosphere. Because glycolysis produces two 3-carbon molecules from glucose, the Krebs cycle occurs twice for every molecule of glucose that enters cellular respiration. Thus, two rounds of the Krebs cycle produce six molecules of CO_2 — all six of the carbon atoms that were originally part of the glucose molecule. The last step of the Krebs cycle rearranges the atoms in the intermediate malate to recreate the intermediate, oxaloacetate, which is needed for the cycle to repeat again.

- **Enzymes transfer energy to ATP.** During one step of the Krebs cycle, energy is transferred to ATP by substrate-level phosphorylation. Because the Krebs cycle occurs twice for every glucose molecule that goes through cellular respiration, a total of 2 ATP per glucose are made during the Krebs cycle.

More is better

Glycolysis only partially breaks down food molecules — the pyruvate from glycolysis still has lots of stored energy that a cell could use! To get that energy, cells move pyruvate into the mitochondrion. As pyruvate enters the mitochondrion, enzymes convert it to a molecule called acetyl-coA.

The conversion of pyruvate to acetyl-coA is called *pyruvate oxidation:*

- **An enzyme oxidizes pyruvate, transferring electrons to NAD^+.** The reduction of NAD^+ forms $NADH + H^+$.

- **An enzyme removes carbon from pyruvate, releasing the carbon as CO_2.** Pyruvate has three carbon atoms, so this reaction produces a two carbon molecule in addition to CO_2. The removal of a carbon atom from an intermediate is called *decarboxylation.*

- **An enzyme attaches a coenzyme, called coenzyme A, to the intermediate, forming acetyl-coA.** Coenzymes help enzymes function correctly. Coenzyme A attaches to the two carbon molecule and helps catalyze the entry of acetyl-coA into the Krebs cycle. As acetyl-coA enters the Krebs cycle, an enzyme removes coenzyme A and releases it back into the mitochondrion.

The conversion of pyruvate to acetyl-coA has many names. Scientists call it pyruvate oxidation, the linking step (because it links glycolysis and Krebs), and the grooming step (because it grooms pyruvate for the Krebs cycle). I prefer the term pyruvate oxidation because I think that term is the most clear about what is actually happening.

Acetyl-coA still contains plenty of stored energy that a cell can use. Cells tap into that energy by sending acetyl-coA through the Krebs cycle. The Krebs cycle is basically a series of redox reactions that transfer energy and electrons from the acetyl-coA to energy and electron carriers.

One of the most important functions of cellular respiration is to transfer energy from food molecules to a form that cells can use to do work. To get the energy out of food, cells oxidize food molecules. So, if you're a cell that is trying to get energy out of food, the more oxidation you do, the better, right? With that in mind, compare glycolysis and the Krebs cycle (Figures 8-5 and 8-6). Glycolysis has one oxidation step, which you can recognize because $NADH + H^+$ is formed. Pyruvate oxidation and the Krebs cycle have a total of five oxidation steps, which you can recognize because $NADH + H^+$ and $FADH_2$ are formed. When you consider that the Krebs cycle happens twice per glucose molecule, that doubles the number of oxidation steps to 10! So, if oxidation of food molecules makes energy available to the cell, it's pretty clear which pathway gives more bang for each glucose molecule — cells that have the enzymes to do the Krebs cycle are much more efficient at extracting energy from their food than are cells that can do only glycolysis!

Chapter 8: Cellular Respiration: Making Your Cake and Eating It, Too

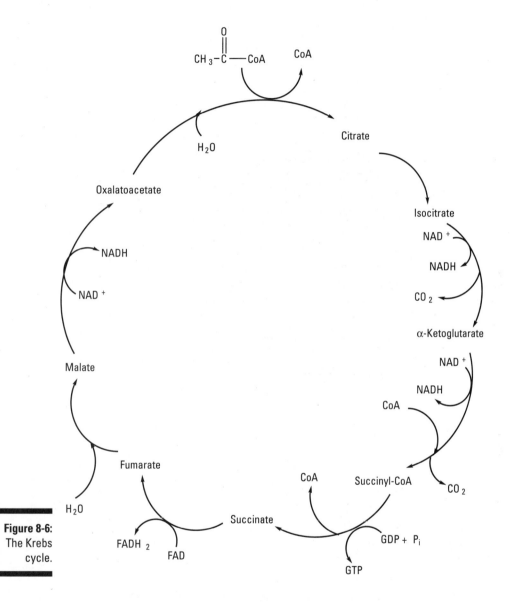

Figure 8-6: The Krebs cycle.

The steps of the Kreb's cycle

The Krebs cycle officially begins with the joining of acetyl-CoA to a four carbon molecule called oxaloacetate. Figure 8-6 shows the details of the reactions in the Krebs cycle:

1. **Enzymes join the 2-carbon molecule acetyl-CoA with the 4-carbon molecule oxaloacetate, creating the 6-carbon molecule citrate.**

   During this process, an enzyme incorporates a water molecule into the intermediate and releases coenzyme-A back to the cell.

2. **An enzyme rearranges the bonds in citrate, creating the molecule isocitrate.**

   The number of carbon atoms in the intermediate doesn't change.

3. **Enzymes oxidize and decarboxylate isocitrate, converting it to the 5-carbon molecule a-ketoglutarate.**

   Enzymes transfer the electrons removed from isocitrate to the electron carrier NAD^+, reducing it to $NADH + H^+$. Enzymes release the carbon removed from isocitrate as CO_2.

4. **Enzymes oxidize and decarboxylate α–ketoglutarate, converting it to the 4-carbon molecule succinyl-CoA.**

   Enzymes transfer the electrons from α-ketoglutarate to NAD^+, reducing it to $NADH + H^+$. Enzymes release the carbon removed from α-ketoglutarate as carbon dioxide CO_2.

   Enzymes require the help of coenzyme A during these reactions and add it to the intermediate.

5. **Enzymes remove coenzyme A from succinyl-CoA and release it back to the cell, changing succinyl-CoA into succinate.**

   This reaction makes energy available, allowing the phosphorylation of ADP by substrate-level phosphorylation and producing a molecule of ATP.

   Substrate-level phosphorylation during the Krebs cycle may produce ATP, or it may produce a very similar energy carrier called guanosine triphosphate (GTP). Two different enzymes can catalyze substrate-level phosphorylation during the Krebs cycle, and each one prefers either ADP or GDP in its active site.

6. **Succinate is oxidized, converting it to fumarate.**

   An enzyme transfers the electrons from succinate to the electron carrier FAD, reducing to $FADH_2$.

7. **An enzyme catalyzes the addition of a water molecule to fumarate, converting it to malate.**

8. **Malate is oxidized, converting it to oxaloacetate.**

   An enzyme transfers electrons from malate to NAD^+, reducing it to $NADH + H^+$.

   Malate oxidation recreates oxaloacetate, the four carbon molecule that's needed at the beginning of the Krebs cycle to join with acetyl-coA. So, as the cell continues to break down food molecules, the Krebs cycle can repeat again and again.

Plants use pyruvate oxidation and the Krebs cycle to break down food molecules — and notice that both of these pathways produce CO_2! Most people think that plants take in CO_2 and release O_2, while humans breathe in O_2 and release CO_2 — which is actually true, but it's not the whole story! Plants do release O_2 and take in CO_2, but they also do what we do, too — they take in O_2 for cellular respiration and release CO_2 from the breakdown of food. It's just that in the light, plants are usually doing way more photosynthesis than respiration, so they're producing much more O_2 than they're using, and consuming way more CO_2 than they're releasing. So, if you measured the gases around a plant in the light, you'd get what everyone expects — overall, the plant would be taking in CO_2 and releasing O_2. However, if you put a plant in the dark, you'd be able to detect the rest of the story — a plant in the dark acts just like a human, taking in O_2 and releasing CO_2. Plants in the dark can't do photosynthesis, so you can actually see the evidence that they do cellular respiration.

Making Useful Energy: Chemiosmosis and Oxidative Phosphorylation

Plants use cellular respiration to break down food molecules so that the plants can access stored matter and energy. For every 6-carbon glucose molecule that's broken down through glycolysis and the Kreb's cycle, plants produce

- Six molecules of CO_2
- Four molecules of ATP
- Ten molecules of NADH + H$^+$, and
- Two molecules of $FADH_2$

By the end of glycolysis and the Krebs cycle, most of the energy from food molecules is stored in the NADH and $FADH_2$ molecules. To transfer that energy to ATP and complete the process of cellular respiration, cells use a process called *oxidative phosphorylation*. Oxidative phosphorylation uses an electron transport chain in the inner membrane of the mitochondrion to transfer energy from reduced electron carriers to ATP by the process of chemiosmosis. (For more details on chemiosmosis, see Chapter 7.)

The *chemiosmotic theory of oxidative phosphorylation* is a really big name for the explanation of how cells use an electron transport chain in the inner membrane of the mitochondrion to transfer energy from reduced electron carriers (NADH + H$^+$) to ATP.

The cost of advertising

Plants are pretty much stuck in one place, so reproduction with other plants presents some challenges. One way that plants get around that challenge is to recruit pollinators to travel from plant to plant. To lure those pollinators in, plants put out some advertising of the smelly variety. First, plants need to make the right perfume for the job, and then they need to get their advertising out there.

To spread the scented word far and wide, plants often rely upon a little warmth. If you've ever lit a scented candle or warmed scented oils, you know that heat helps scent molecules travel through the air. The heat gives the scent molecules more kinetic energy, causing them to vibrate faster and disperse more quickly. So, when plants want to put out a scent, a little heat gets the job done faster.

How do plants generate the heat? Cellular respiration! It turns out that whenever energy is transferred by cells, the process isn't perfectly efficient. Only about 40 percent of the energy from food gets transferred to ATP during cellular respiration; the other 60 percent becomes heat energy that's transferred to the environment. So, when plants want to turn up the heat, they crank up the cellular respiration. Heat transfers through the plant tissues, helping spread their scent molecules and attracting pollinators.

For comparison, think about yourself when you exercise. Your cells turn up the cellular respiration to make more ATP for your muscles, and your body temperature goes up. (You also typically become more, er, fragrant.)

Transferring electrons

NADH and $FADH_2$ donate electrons to the electron transport chain during oxidative phosphorylation.

The electrons are pulled from complex to complex in the chain until they are accepted by O_2 at the end of the chain. When oxygen is reduced, it also accepts some H^+, forming H_2O.

Molecular oxygen (O_2) is the final electron acceptor during oxidative phosphorylation in plants. Oxygen is very electronegative, so it makes a great electron acceptor.

NADH and $FADH_2$ are oxidized as they donate electrons to the beginning of the electron transport chain, converting them back to NAD^+ and FAD. The oxidized carriers can now go back to glycolysis and Krebs and accept more electrons.

Transferring energy

As electrons move through the electron transport chain during oxidative phosphorylation, energy is transferred to ATP via chemiosmosis.

The essential steps of chemiosmosis during oxidative phosphorylation are the same as those that occur during chemiosmosis of photosynthesis:

1. **Energy and electrons are donated to an electron transport chain.**

   During cellular respiration, the energy and electrons originally came from food and are donated to the chain by NADH + H⁺ and FADH$_2$.

2. **As the electrons move through the chain, the protein complexes access some energy to pump protons (H⁺) across a membrane.**

   The concentrated hydrogen ions represent a source of potential energy called proton motive force.

   During oxidative phosphorylation, the protons are pumped across the inner membrane of the mitochondrion, into the intermembrane space between the two mitochondrial membranes.

4. **The protein ATP synthase allows hydrogen ions to move back across the membrane.**

   As the hydrogen ions move, they cause part of ATP synthase to rotate, which increases energy in the ATP synthase. The enzyme uses this energy to join ADP and phosphate groups, making ATP. ATP finally stores the energy from food!

Plants can now use the ATP as a source of energy for growth, repair, movement of molecules, and other types of cellular work.

ATP synthase is embedded in the inner mitochondrial membrane right along with the electron transport chain (see Figure 8-7). On one side of the membrane is the intermembrane space; on the other side is the matrix of the mitochondrion. During oxidative phosphorylation, hydrogen ions (H⁺), which are also called protons, enter a small channel in ATP synthase from the intermembrane space side of the membrane. The protons bind to the ATP synthase and then exit the protein on the matrix side. As they bind to the ATP synthase, the head of the protein rotates. The base of the protein has binding sites for ADP and inorganic phosphate (P). The passage of every three protons through the ATP synthase provides enough energy for the synthesis of one ATP molecule from ADP and P.

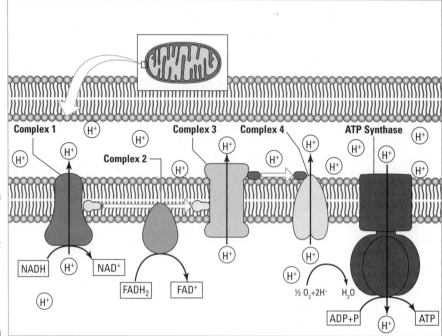

Figure 8-7: The electron transport chain in the inner membrane of the mitochondrion.

Chemiosmosis occurs as part of cellular respiration and also photosynthesis (see Chapter 7).

Chapter 9
Moving Materials Through Plants

In This Chapter
- Transporting molecules across plasma membranes
- Making water flow according to the cohesion-tension theory
- Moving sugars with the pressure-flow hypothesis

*P*lants make sugars in their leaves and then must move those sugars throughout the plant in order to provide sugar to all the plant cells. Likewise, water and minerals are taken up by plant roots and then must be transported to all the plant cells. Once the circulating materials arrive at plant cells, they must be moved into the cells. This chapter explains how plants move sugar and water throughout their bodies and how individual plant cells get what they need to survive.

Shipping and Receiving Materials throughout the Plant

You can think of a plant like a factory, with different areas that specialize in the manufacturing of components that everybody needs. The leaves specialize in making food, while the roots specialize in getting water and minerals. However, every cell in the plant needs both of these components in order to survive. So, plants need efficient shipping mechanisms to circulate both sugar and water throughout their bodies. Figure 9-1 gives an overview of how these materials circulate through plants so that every cell can get what it needs.

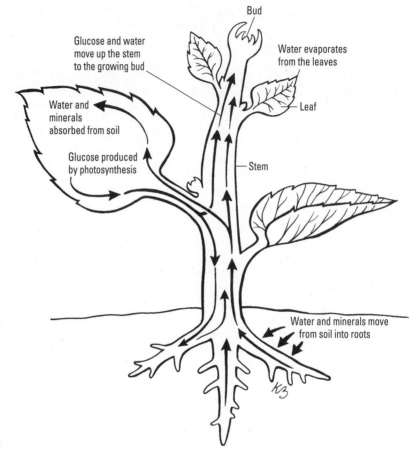

Figure 9-1: Overview of the movement of molecules through plants.

Moving Across Membranes

The plasma membrane controls what can enter and exit each plant cell. (See Chapter 2 for the basic structure and function of plasma membranes.) Whether or not a molecule will cross a membrane, and which way it will go, depends upon three things:

- **The chemical structure of the molecule, particularly its size and whether it's hydrophobic or hydrophilic.** Small, hydrophobic molecules can cross the lipids in the plasma membrane all by themselves. As molecules get larger, or if they're hydrophilic, they can cross the membrane only if a transport protein in the membrane helps them across.

- **The concentration of the molecule on either side of the membrane.** If molecules are allowed to move freely, they will move from areas where they're more concentrated to areas where they are less concentrated — in other words, they'll move until they're evenly spread out, or in *equilibrium,* over an area. This type of movement, called *diffusion,* is explained in more detail in the next section.

✔ **Whether or not the membrane contains a transport protein for the molecule.** Although some small hydrophobic molecules can cross membranes by themselves, most molecules need the help of a transport protein. If a plant plasma membrane doesn't have a protein to transport a particular molecule, it won't be able to cross in or out of the plant cell.

Diffusion

Diffusion is the movement of molecules from an area of higher concentration to an area of lower concentration. Diffusion is a passive process and requires no energy to be input from the cell.

Diffusion happens because of the *kinetic energy,* or energy of motion, of molecules themselves. Because of kinetic energy, molecules are constantly in motion and jiggle and bounce themselves around, eventually becoming randomly distributed throughout an area.

Brownian motion is the random movement of molecules due to kinetic energy.

Figure 9-2 shows how diffusion across a membrane works:

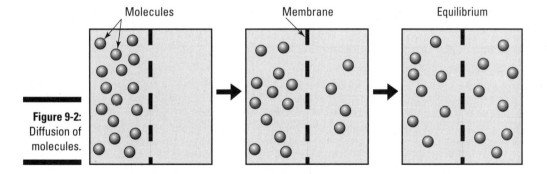

Figure 9-2: Diffusion of molecules.

1. **A chemical starts out more concentrated on one side of a membrane, but its molecules can cross the membrane freely.**

2. **The molecules bounce around due to kinetic energy and gradually become equal in concentration on both sides of the membrane.**

3. **If you measure the overall movement of the chemical, you can see that its molecules moved from the side of the membrane where they were more concentrated to the side where they were less concentrated.**

When molecules move from where they are more concentrated to where they are less concentrated, scientists say that molecules are moving along their *concentration gradient*.

Figure 9-3 shows two basic types of diffusion that occur across plant cell membranes:

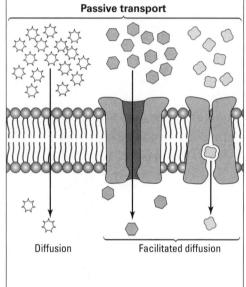
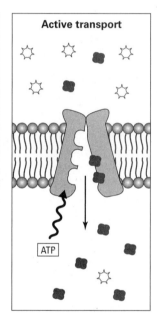

Figure 9-3: Transport across membranes.

- *Simple diffusion* occurs when molecules diffuse across a membrane all by themselves.
- *Facilitated diffusion* occurs when molecules diffuse across a membrane with the help of a membrane protein. (See Chapter 2 for details on how proteins help molecules across membranes.)

Even though facilitated diffusion involves the help of a protein, the movement of molecules still occurs from an area of higher concentration to an area of lower concentration. So, the process is still diffusion and doesn't require any input of energy from the cell.

Active transport

Plant cells often need to move molecules from areas where the molecules are less concentrated to areas where they're more concentrated, a process called *active transport*. In other words, plant cells need to move molecules against their concentration gradient.

Active transport requires the input of energy from the cell. Active transport is like rolling molecules up the hill of their concentration, going from where they're less piled up to where they're more piled up. If a cell is going to move something up hill, it's going to have to put some energy into the process.

Figure 9-3, earlier in this chapter, shows how a carrier protein could move a molecule against its concentration gradient by getting energy from ATP molecules during the transport process. Scientists call these proteins *active transport proteins,* or *pumps.* As an example, a plant cell that stores energy for the plant may have lots of glucose inside the cell, but it may still need to bring in more. This cell could use active transport to continue bringing glucose molecules in against their concentration gradient.

Osmosis

Water moves across membranes by diffusion. The diffusion of water across membranes is called *osmosis.* During osmosis, water moves from an area where water is more concentrated to an area where water is less concentrated.

Thinking about osmosis in terms of water concentration can be a little hard to wrap your head around: Water is more concentrated when fewer solutes are dissolved in the water. In other words, water is most concentrated where water is most pure. So water moves from where water is most pure to areas where water is less pure.

If thinking of osmosis in terms of water concentration is confusing, think of osmosis in terms of solute concentration instead. Water will diffuse into an area of greater solutes. The concentration of solutes on either side of a membrane can be described by three terms that are used to describe the relative concentrations of solutes in different solutions:

- *Hypertonic* solutions have greater concentrations of solutes (*hyper-* means more).
- *Hypotonic* solutions have lesser concentrations of solutes (*hypo-* means under).
- *Isotonic* solutions have the same concentration of solutes (*iso-* means same).

If you have two different concentrations of a chemical on either side of a membrane, water will move by osmosis toward the hypertonic solution. As an example, refer to Figure 9-2. Solutes move by diffusion from left to right until they're evenly distributed across the membrane. But, in that same scenario, what would the water around those solutes be doing?

1. **Initially, the solution on the left side of the membrane is hypertonic to the solution on the right side of the membrane because more molecules are dissolved in the solution on the left side.**

2. Water molecules move by osmosis toward the hypertonic solution — in other words from the right side to the left side of the membrane.

3. If the membrane is permeable to both the solute and the water, they will diffuse in opposite directions until both are randomly distributed across the membrane.

Living cells don't allow everything to pass freely across their membranes. Certain solutes require a protein transporter to help them get across. So, in a living cell, water may move by osmosis until its concentration is the same across a membrane, while a particular solute may remain unevenly distributed. To determine whether the solute will move, you need to know whether it needs a transport protein and, if so, whether the cell in question has that transport protein.

Whether you think of osmosis in terms of water concentration or movement toward solutes, the result is the same: Water molecules become more randomly distributed by osmosis.

Under pressure

Water movement in plant cells is complicated by the fact that plant cells may also be under pressure. As plant cells fill up their central vacuoles with water, their plasma membranes push against the walls around the cells. The pressure on plant cells that results from the uptake of water is called *turgor pressure*.

As water flows into a plant cell and the cell becomes increasingly firm, or *turgid,* the increase in turgor pressure may slow the movement of water even if the water concentration hasn't yet reached equilibrium.

To understand the relationship between osmosis and turgor pressure, think of a plant cell as a water balloon inside a shoe box. In this case, the latex of the balloon represents the plasma membrane of the cell, and the box is the cell wall. The water balloon is filled with salty water to represent the cytoplasm of the cell. If you attach a hose to the water balloon so that pure water comes into contact with the cytoplasm, then water will flow into the balloon. The water is moving by osmosis from an area of low solutes (pure water) to an area of high solutes (the salty solution inside the balloon). As water moves into the balloon, however, the balloon stretches and begins to push against the box. Eventually, the pressure from the box will prevent any more water from flowing into the balloon, even if the solution inside the balloon hasn't reached equilibrium with the solution from the hose.

Whether or not water moves in or out of plant cells depends upon both turgor pressure and solute concentration. Botanists combine both of these factors to determine *water potential* of a cell. Water potential is represented by the Ψ symbol and is measured in megapascals (mPa), which is a unit of pressure (one atmosphere, which is the pressure that the atmosphere puts on the earth at sea level, equals 0.1 mPa).

You can think of water potential as the tendency of water to leave one place in favor of another. To figure out the water potential of a cell or solution within a plant, you add together the effects of solute concentration and turgor pressure:

- **Botanists express the effect of solute concentration as *solute potential* (Ψ_S), which is sometimes called *osmotic potential*.** The solute potential of pure water equals zero. As solutes are added to water, the solute potential becomes more negative, so plant cells always have negative solute potential.

- **Botanists express the effect of turgor pressure as *pressure potential* (Ψ_P).** Normal atmospheric pressure at sea level has a pressure potential equal to 0. As turgor pressure in a cell increases above normal atmospheric pressure, pressure potential becomes more positive.

To determine overall water potential, solute potential and pressure potential are added together using the equation:

$$\Psi = \Psi_S + \Psi_P$$

Water always moves from an area of higher water potential to an area of lower water potential.

Table 9-1 shows how the water potential equation compares to the concepts of how water moves in response to solute concentration and pressure.

Table 9-1	Relating Concepts of Water Movement to the Water Potential Equation
Concept	**Equation**
Water moves by osmosis to an area of greater solute concentrations.	As solute concentration increases, the value of Ψ_S decreases, lowering the overall water potential. So, as solute is added to a solution, its water potential decreases, which increases the chances that water will move into the solution.
Increasing turgor pressure will slow the movement of water into a plant cell.	As turgor pressure increases, the value of Ψ_P increases, raising the overall water potential. So, as turgor pressure increases, the overall water potential increases, which lessens the chances that water will move into the cell.

Losing it

Water can move into plant cells, and it can move out of plant cells, too. If you put crisp celery or carrot sticks into a salty cup of water, they'll pretty quickly get limp and rubbery as they lose their turgor pressure. Because of the dissolved salt, the salty water has a lower water potential than the plant cells. Water flows out of the plant cells, and the cells no longer push up against their cell walls. Turgor pressure is lost, and the vegetables become limp.

If you looked at the cells in the vegetables under a microscope, you'd see something like the *plasmolyzed* cell drawn in Figure 9-4. When water leaves a walled cell through osmosis, the cytoplasm collapses, and the plasma membrane pulls away from the cell wall. This collapse of the cell is called *plasmolysis*.

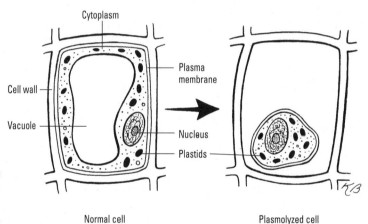

Figure 9-4: Plasmolysis.

Going with the Flow: Water Transport

Plants transport water through their *xylem*, which is mostly made up of dead cells whose strong secondary cell walls form tubes throughout the plant. (See Chapter 3 for details on the cells that make up this tissue.) So, the xylem itself isn't actively moving water through the plant; instead, it's like the pipes in the plumbing of your home — hollowed out tubes through which water can move.

Water moves through plants by a combination of two factors:

- ✔ The adhesive and cohesive properties of water
- ✔ The pull of water evaporation from the leaves *(transpiration)*

Botanists call the explanation for how water moves through plants the *Cohesion-Tension Theory*.

Clinging for support: The cohesion of water

Water molecules stick together due to the hydrogen bonds that form between them. Although each individual hydrogen bond is just a weak electrical attraction, when you combine lots of hydrogen bonds together, the attraction becomes very powerful — so powerful, for example, that a person can put on a pair of water skis and slide across the surface of a lake. If water molecules didn't stick together to create a resistance called *surface tension,* the skier would fall right in!

Cohesion describes the attraction between water molecules due to their hydrogen bonds. *Adhesion* describes the attraction between water and other substances.

One way to visualize the importance of cohesion and adhesion in water movement through xylem is to look at how water moves through narrow tubes called *capillary tubes. Capillary action* occurs when a narrow tube pulls a liquid upward against the force of gravity. Figure 9-5 illustrates capillary action in tubes of different diameters:

- **When narrow tubes are placed in water, water will automatically creep upward in the tube due to adhesive forces between water molecules and the molecules that make up the tube.** The water molecules are attracted to the molecules in the solid material, such as glass, that makes up the tube. The adhesive forces between the water molecules and the sides of the tube cause the water molecules to cling to the solid surface. The water molecules along the edges cling to the solid tube and also to the other water molecules, pulling the water upward in the tube.

- **The narrower the tube, the higher the water will move.** Gravity also acts on the water molecules, pulling them downward. So, the height of the column of water depends upon the balance of forces between the attraction to move upward in the tube and the downward pull of gravity. In a narrow tube, more contact occurs between the water molecules and the molecules of the tube itself, so relatively more force is causing the water to move upward than there would be in a wider tube.

In a plant, capillary action occurs when water moves through the narrow tubes of xylem:

- Cohesive forces between the water molecules themselves act as the glue that holds the column of water together.

- Adhesive forces between the water molecules and the cellulose microfibrils that make up the cell walls of the xylem draw the water column up the xylem. The water column will rise until the adhesive forces are balanced by the downward pull of gravity. Xylem tubes are very narrow, so lots of contact occurs between the water and the cell walls, helping the upward movement of the water column.

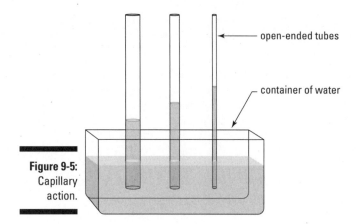

Figure 9-5: Capillary action.

The pull from above: Transpiration

Capillary action (see preceding section) explains how a column of water would rise upward and then remain hanging in the narrow tubes of xylem in a plant, but it's not enough to explain what makes water keep flowing upward from the roots to the leaves. Water moves upward through the xylem because water that evaporates from the leaves is replaced by water from the continuous column in the xylem. The evaporation of water from the leaves is called *transpiration*.

Water moves through a plant and out into the atmosphere because of the gradient of water potential created by transpiration. (See the earlier section "Under pressure" for an explanation of water potential.) Figure 9-6 shows the flow of water through a plant and out into the atmosphere:

1. **The atmosphere has a very negative water potential, more negative than the water potential of the air in the leaf.**

   So, water moves from the leaf air into the atmosphere.

2. **The leaf air has a more negative water potential than the leaf cells, so water moves from the leaf cells into the leaf air.**

3. **The leaf cells have a more negative water potential than the column of water in the xylem, so water moves from the xylem into the leaf cells.**

4. **The column of water has a more negative water potential than the root cells, so water moves from the root cells into the xylem.**

5. **The root cells have a more negative water potential than the soil, so water moves from the soil into the root cells.**

The rate of water movement through the plant depends upon the rate of transpiration. Anything that increases the rate of transpiration will increase the rate of water movement through the plant.

Chapter 9: Moving Materials Through Plants

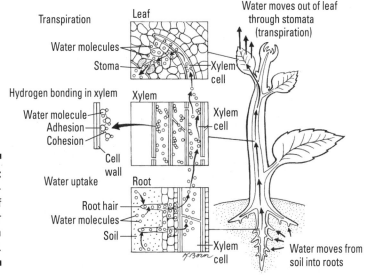

Figure 9-6: The movement of water through a plant.

Several environmental factors can increase the rate of transpiration:

- When the air around a plant is dry, its water potential is lower, increasing the rate of transpiration.
- If the wind is blowing, water vapor surrounding the leaf will be swept away, which also lowers the water potential of the air around the leaf, increasing transpiration.
- Increases in temperature increase the vapor pressure of water, which increases the rate of transpiration.
- Anything that causes stomata to open has the potential to increase transpiration.

Different signals cause stomata to open:

1. The guard cells that surround the stomata have receptors for environmental signals, such as light and carbon dioxide (CO_2).
2. When the guard cells detect that light is available or that the concentration of CO_2 is low, they actively transport hydrogen ions (H^+) out of the cell, causing the guard cells to become negatively charged.
3. The negative charge on the guard cells encourages positively charged potassium ions (K^+) to enter the cell, decreasing the water potential of the guard cells (because solute potential becomes lower).
4. When the water potential of the guard cells is lowered, water moves by osmosis into the cells, making them more turgid.
5. When guard cells become turgid, they pull away from the stomata, allowing more transpiration to occur.

Transpiration essentially pulls the water columns up through the plant. The pulling force of transpiration is the opposite of the pushing force of pressure, so botanists refer to it as *negative pressure* or *tension*.

Fixing a break in the lines: Cavitation

In order for water columns to move up plants, the water molecules need to stick to each other. If anything happens to introduce an air bubble into the column of water, the column of water below the air bubble will drop, and water movement in that column will stop.

The sudden formation of gas bubbles in a liquid is called *cavitation*. Several factors can cause cavitation in the xylem of plants:

- On a hot, dry day, the tension on water columns due to rapid transpiration may exceed the adhesive forces that hold the water column together. The water molecules get pulled apart and separated by a gas bubble.
- Dissolved gases in the water column may come out of solution, creating a gas bubble.
- Freezing and thawing cause changes in the arrangements of the water molecules that can lead to the introduction of gas bubbles.

Because plants rely upon cohesion of water molecules to move water through the plant, cavitation is potentially a serious problem. But, as is typical of living things, where there's a problem, there's a solution:

- **Plants that experience high levels of cavitation may rely upon tracheids rather than vessels for water transport.** Vessel elements are wide open tubes, whereas tracheids have more enclosed ends. Also, tracheids have pits that allow for lateral water transport as well as vertical water transport. So, if an air bubble occurs in a vessel element, the whole water column is blocked. But if an air bubble occurs in a tracheid, the bubble may be contained to a single tracheid cell. Water may be able to find a new path around that blocked cell by traveling through the pits of the surrounding tracheids. Arctic plants, which experience lots of freezing and thawing that can cause cavitation, rely upon tracheids instead of vessels.
- **Woody plants form new xylem every year.** One solution to xylem that doesn't work anymore because of air bubbles is to just ignore it and make new xylem.

- **At night, plants use *root pressure* to force water upward and repair broken water columns.** At night, root cells release ions into the xylem, lowering its solute potential and thus its water potential. As a result, water flows into the xylem. If the incoming water pushes a broken water column up through an air gap until it touches the rest of the column, then the water column is restored. Root pressure is most effective in restoring water in shorter plants as the maximum height a water column can be forced upward by this method is about 60 feet.

Reaching out with roots

Water enters plants through the roots. As rain enters the soil, the water molecules are attracted to soil particles. The water coats the soil particles, forming multiple layers of water shells. The water shells closest to the soil particle have the most negative water potential, so the innermost water shells are held tightly to the soil particle. In order for a plant to get water from the soil, root cells must come into contact with the water shells around the soil particle. Because water always moves to the lower water potential, plants can attract water only from the shells that have a higher water potential than the root cells themselves. So, roots typically can't take all the water from the soil.

In order to get enough water to survive, plants increase contact between root cells and soil particles (see Chapter 4 for details on root structure, root hairs, and mycorrhizal fungi):

- **Branch roots** spread more root mass through the soil, increasing the contact between roots and soil particles.
- **Root hairs** extend out into the soil, increasing the surface area of contact between root cells and soil particles.
- **Tap roots** extend down into the soil in order to contact water that maybe found deep in the soil.
- **Mycorrhizal fungi** that associate with roots also extend the reach of roots throughout the soil.

As roots take up water, they also take up *mineral nutrients* for plant growth. Plants get 13 minerals from the soil (see Chapter 7). As rocks in the soil dissolve, they release ions into the soil. These ions are dissolved in the water shells around the soil particles and get taken into plants as they take up water. It's not always easy for plants to get these minerals because positively charged ions, called *cations,* are attracted and cling to the negatively charged surfaces of clay particles in the soil. But plants can loosen these minerals up by doing *cation exchange* — plants release cations, such as hydrogen ions, (H^+), which are also attracted to the clay and can bump off some of the other cations, making them available for absorption by the plant. Because ions can only enter plant cells through membrane transport proteins, cells in the plant roots screen the ions and only allow valuable ions to enter the plant. And, the waxy *casparian strip* (see Chapter 4) helps make sure that all water flowing through the root must enter root cells at some point.

Sticky Business: Sugar Transport

Plants transport sugars through their phloem. Like the xylem, phloem is made up of tubes of cells connected end to end. However, unlike the cells of the xylem, the sieve tubes and companion cells of phloem aren't dead. The cells of phloem transport ions and sugars across their membranes in order to start the process of sugar transport.

Plants move sugar through their phloem by actively transporting solutes in order to manipulate the flow of water. Turgor pressure in the phloem is increased in areas of the plant where sugars are available *(sources)* and decreased in areas of the plant where sugars are needed *(sinks)*.

Sources and sinks

Green parts of plants, such as leaves, make sugars through photosynthesis, so they're the ultimate source of sugars for plants. In addition, sugars may be stored in roots or stems and then accessed later when needed. When sugars are taken from the roots — for example, in the spring — then roots become a source for sugars for the plant.

Every part of the plant needs sugar, especially rapidly growing areas like stem tips and developing fruits. These high demand areas are sinks for sugar. During the growing season, plants may use roots or underground stems for storage, making them another sink for sugar. And in the spring, when new buds break open and begin to grow, they become a significant sink for sugars.

Pressure-flow hypothesis

Botanists studying the movement of sugars through phloem have summed up their observations in an explanation called the *pressure-flow hypothesis*. According to this hypothesis, shown in Figure 9-7, sugars are moved through phloem by the movement of water in response to changes in water potential that result from changes in solute potential and pressure potential.

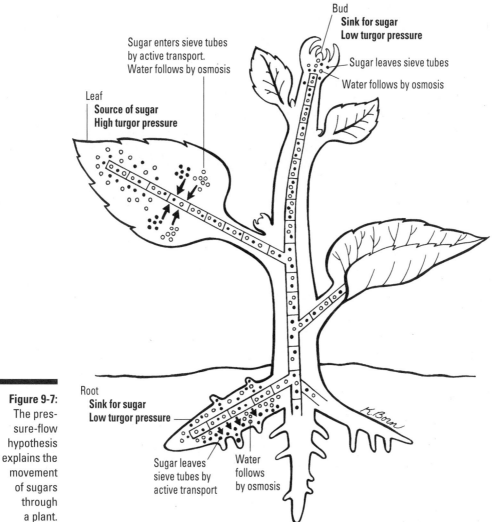

Figure 9-7: The pressure-flow hypothesis explains the movement of sugars through a plant.

Events at the source are

1. **Phloem cells at the source actively transport sugars into the sieve tubes.**

2. **As sugars enter the sieve tube, the solute potential becomes more negative, which lowers the water potential.**

3. **Water enters the sieve tube from nearby xylem by osmosis because of the lowered water potential.**

4. **The influx of water increases pressure in the sieve tube and pushes material through the tube.**

   Because sieve tube cells have porous cell walls at their ends, much of the fluid part of each cell is moved to the next cell in a process called *bulk flow*.

Events at the sink are

1. **Plant cells at the sink remove sugars from the sieve tubes by active transport and either use the sugars immediately or convert them into starch for storage.**

2. **As sugars leave the sieve tube, the water potential inside the sieve tube becomes less negative.**

3. **Water leaves the sieve tubes by osmosis and moves into the surrounding tissues which have a lower potential.**

4. **As water leaves the sieve tubes, the pressure inside the tubes decreases.**

The pressure gradient between sources and sinks causes sugars to move through the phloem from sources to sinks.

Chapter 10
Regulating Plant Growth and Development

In This Chapter
- Comparing types of plant growth
- Looking at signal transduction
- Exploring the effects of plant hormones
- Discovering how plants measure light with phytochrome

*P*lants grow through cell division and cell elongation, and then cells differentiate to form the specialized cells, tissues, and organs of the plant. Plant growth and development is regulated by the production of hormones. Plants time their growth and reproduction by using receptors to perceive environmental signals such as light. This chapter explores how plant hormones and photoreceptors are used to regulate plant growth and development.

Getting an Overview of Plant Growth and Development

As plants grow, they increase in height and weight. While growth results in increased size of the plant, *differentiation* results in new or more specialized functions of plant structures.

At the cellular level, plant growth results from two different processes:

- **Division growth:** Cell division by meristems produces growth by the addition of new cells.
- **Elongation growth:** Cell elongation produces growth as single cells expand.

Growth that increases the height of plants occurs at the tips of shoots and roots. At the very tip of both roots and shoots, apical meristems produce new cells, creating organized tissues. Just behind the meristematic tissue of both shoots and roots is the elongation zone where most of the elongation growth occurs.

As cells elongate, they also mature and become specialized for different functions.

As cells mature, they differentiate and change their structure to suit their function.

Once cells have differentiated and are mature, they typically no longer grow.

The behavior of plant cells — whether they divide, elongate, or differentiate — is controlled by plant hormones, which are released in response to the state of the plant's environment.

Hormones are chemicals produced in tiny amounts in one part of an organism that then travel to another part of the organism where they cause a change in function of the target cell.

Receiving signals

In order for a cell to respond to a hormone, it must have a *receptor* for that hormone. Receptors are proteins that bind very specifically to a certain signaling molecule, like a hormone. Once a hormone binds to the receptor, the receptor relays the signal to the inside of the cell in a process called *signal transduction*.

These three main steps occur during signal transduction in a plant cell:

1. **Reception: A hormone binds to a receptor in the plasma membrane of a cell.**

2. **Signal transduction: The part of the receptor that extends into the cell changes, activating molecules inside the cell called *second messengers*.**

3. **Induction: Second messengers cause a change in the behavior (based on chemical reactions) of the cell.**

Responding to signals

Cells respond to signals like hormones by changing their behavior: They may produce new structures, change their metabolism, or start the process of cell division. Each change requires the cell to use certain proteins and not others.

Chapter 10: Regulating Plant Growth and Development

So, in order for a cell to change its behavior, it must make the right proteins for the new task at hand. The instructions for making proteins are written in the genetic code of DNA and contained in sections of chromosomes called *genes*.

When cells need to make a new protein for a particular task, they copy the instruction in the gene for that protein and then make the protein using the cell's ribosomes. So, if a cell receives a signal to change its behavior, it can respond by accessing new genes and building new proteins.

Gene regulation is the process that causes cells to access certain genes and not others — or as scientists say, to turn genes on and off. Ultimately, signals like hormones cause changes in behavior because they cause changes in gene regulation. After the hormone binds to its receptor, the activated second messengers cause genes to be turned on or off in the cell. Thus, hormones control growth and differentiation of plant cells. *Development* is the process of growth and differentiation of cells into tissues, then tissues into organs, and organs into organisms.

Sending Signals with Plant Hormones

Six major groups of hormones control the development of plants. Table 10-1 summarizes the origin and major effects of each of these groups. The following sections add details on how these hormones coordinate specific plant responses.

Table 10-1	Six Major Groups of Plant Hormones	
Hormone Group	**Origin in Plant**	**Effects on Plant**
Auxins	Produced by apical meristems, buds, and young leaves	Phototropism, gravitropism, apical dominance, inhibition of leaf abscission, flower initiation, sex determination, fruit development
Cytokinins	Mainly produced in roots and exported upward into plant	Cell division, cell differentiation, lateral bud development, stimulate cotyledon growth, delay of senescence, opening of stomata

(continued)

Table 10-1 *(continued)*

Hormone Group	Origin in Plant	Effects on Plant
Gibberellins	Produced by apical meristems, young leaves, and embryos	Cell elongation, rapid stem elongation, breaking of dormancy, promotion of flowering
Abscisic acid	Produced in plastids so found in green tissues, also found in fleshy fruits	Growth inhibition, closing of stomata, delay of seed germination
Ethylene	Produced by fruits, flowers, seeds, leaves, and roots	Ripening of fruit, leaf abscission, senescence, initiation of elongation and bud development
Brassinosteroids	Produced in reproductive organs and growing tissues, such as shoots and immature seeds	Cell elongation, cell division, xylem differentiation, pollen tube growth, seed germination, leaf senescence, inhibit lateral bud growth

Auxins

Auxins were the first plant hormones to be discovered, and they influence many aspects of plant growth and development. Auxins were discovered when scientists investigated the ability of oat seedlings to bend toward the light.

A Dutch botanist named Frits Went set out to find out what exactly in the young oat seedlings was causing them to bend toward the light. Figure 10-1 shows the results of his experiments:

1. **Went cut off the tips of young oat seedlings, called *coleoptiles*, and placed the shoot tips on a slab of a gelatin-like material called agar.**

    If the tips of the oat seedlings were making a chemical that influenced shoot growth, that chemical could soak into the agar.

2. **After letting the seedling shoot tips rest on the agar for a few hours, he cut the agar up into little squares and then tested how beheaded seedlings would respond to the agar squares.**

    If the agar squares contained chemicals from the shoot tips, then putting the agar squares on top of a beheaded seedling would be a way of applying these chemicals to the seedlings.

3. **If he put the agar square directly on top of a beheaded seedling, the seedling grew straight up.**

   These seedlings behaved like normal seedlings, which showed that the agar blocks could replace the function of the missing shoot tips. Went used the behavior of these seedlings as a comparison for his other experimental treatments, so this group of seedlings represents the *control group* in his experiment.

4. **If he put the agar square off to one side of a beheaded seedling, the seedling curved away from the agar square.**

   By putting the agar square off to the side, Went was applying the chemicals to just half of the beheaded seedling. This way, he could compare the behavior of the two halves. The cells in the half that were exposed to the chemical grew more quickly than the cells in the other half, causing the side of the seedling under the agar to get longer, faster. This made the shoot tip curve toward the side that didn't get the chemical — in other words, away from the side with the agar square.

Went's experiment showed that the coleoptile tips were making something that could diffuse into the agar and that was causing the tips to bend. Went named the substance *auxin* for a Greek word that means to grow.

Apical dominance

If you've ever pinched the top off a plant in order to encourage the side branches to grow, then you already know something about apical dominance. *Apical dominance* is the tendency of most plants to grow upward more than outward. Apical dominance results from the ability of apical meristems to inhibit lateral buds.

As long as an active apical meristem is located near lateral buds, the lateral buds remain dormant. If you pinch off the apical meristem, then the inhibition disappears, and the lateral buds begin to grow. So, if you want bushier house plants, you can pinch off the ends of the growing branches to encourage side branches to grow. As a plant gets taller, the lower lateral buds get further from the source of auxin, so they're not inhibited as strongly and may start to grow. In trees such as conifers, atypical dominance often leads to a triangular Christmas-tree appearance.

Botanists demonstrated that the inhibitory effect on lateral buds is due to auxin by doing a simple experiment. First, they pinched the tips off stems and showed that the lateral buds began to grow. Then, they pinched the tips off stems and replaced the tips with wax plugs that contained auxin. The lateral buds remained dormant just as if the apical buds were still present.

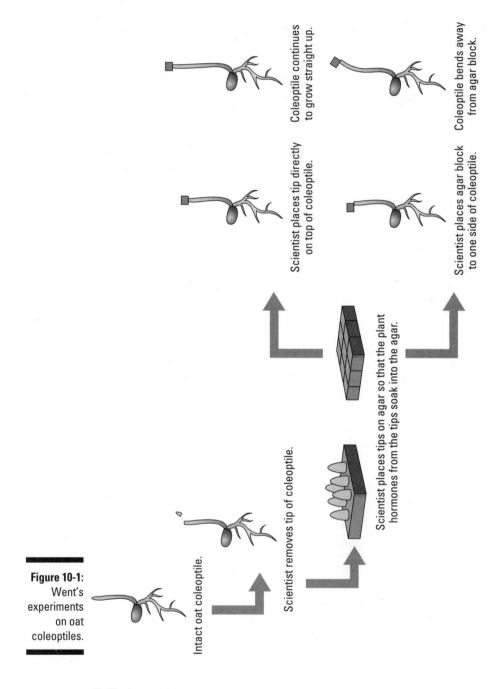

Figure 10-1: Went's experiments on oat coleoptiles.

Cell elongation

Auxin targets cells in the elongation zone of a stem, enabling them to stretch. Botanists demonstrated this effect of auxin by cutting the elongation zone out of stems. If they gave the stem pieces sugar, water, and minerals, they could keep them alive, but the stem pieces wouldn't grow. But when the scientists added auxin to the mix, elongation occurred.

Many botanists believe that auxin plays a role in cell elongation according to a process called the *acid-growth theory:*

1. **Auxin activates a membrane transport protein called ATPase and causes that protein to pump hydrogen ions (H+) into the cell wall.**
2. **As H+ concentration increases, the pH of the cell wall decreases and the cell wall becomes acidic.**
3. **The acidic conditions in the cell wall cause loosening of the bonds that hold the wall together, for example by favoring the activity of enzymes that break bonds in the hemicellulases in the wall.**
4. **Once acidification loosens the cell wall, it's easier for cells to expand.**
5. **Auxin also activates growth genes that are required for sustained cell elongation to occur.**

Cytokinins

Like auxins, cytokinins have many effects on plant growth and development. They were first discovered because of their ability to stimulate cell division and trigger the production of the cork cambium in woody plants.

One of the most interesting aspects of cytokinins is the balancing relationship they have with auxins. Several growth responses in plants depend not just on the presence or absence of these two hormones, but on the ratio of the two.

Botanists have experimented with these responses by growing plant tissues in the laboratory in a process called *tissue culture*. Pieces of mature stem, leaves, or roots are cut from plants, placed on agar and nutrients, and then exposed to different ratios of auxin to cytokinins:

- If the amount of auxin and cytokinin are equal, the stem pieces will grow into a disorganized blob of cells called a *callus*.
- If the amount of auxin is high and the cytokinins are low, then the plant tissue develops roots.
- If the amount of auxin is low and the cytokinins are high, then the plant tissue develops shoots.

Tissue culture is widely used in nurseries to propagate slow-growing plants, such as orchids, and in many botanical experiments requiring many clones of a plant. Tissue culture is also the process used to make many clones of a genetically engineered (transgenic) crop or horticultural plants. Tissue culture demonstrates that the correct ratio of auxins and cytokinins is essential for plants to develop their organs in the right places. If you've ever taken a plant cutting and placed it in a glass of water to get it to root, you've tested this idea.

Auxins are produced primarily in shoot apical meristems, while cytokinins are produced primarily in root apical meristems.

By taking a cutting, you separate a piece of shoot from the roots, creating a little plant that had a high auxin to cytokinin ratio. The imbalance in the hormones signals the plant about what it's missing, and it responds by making new roots!

Gibberellins

Gibberellins are probably best known for their ability to stimulate cell division in the internodes of plants. If you've ever grown lettuce or cabbage in your garden, you've probably seen the effect of gibberellin. In spring, these plants form lots of leaves very close together, forming a rosette or head of leaves. These rosettes form because the internodes between the leaves are very short. But then, as the days get longer and warmer, your cabbages and lettuces may suddenly bolt, rapidly sending up a flowering stem that has long internodes between the leaves. The longer days and warmer weather triggers the production of gibberellins, leading to the change in growth of your garden plants.

Some plants, such as peas and beans, produce dwarf varieties that are short because they completely lack the ability to make this hormone. When gibberellin is applied to these dwarves, their internodes grow rapidly, and they look like their nondwarf relatives. For example, bush beans that you might grow in your garden can be transformed into pole beans by the application of gibberellin.

Abscisic acid

Abscisic acid inhibits plant growth and triggers bud dormancy. In fact, commercial growers apply abscisic acid to nursery plants in order to trigger dormancy before shipping so that buds are less likely to be damaged in transit.

When leaves are under water stress, they increase production of abscisic acid. The abscisic acid interferes with the transport of potassium ions in guard cells, causing them to become flaccid and close the stomata (see Chapter 9). By closing the stomata, the abscisic acid helps protect the plant from further water loss.

Foolish rice

In the 1920s, a Japanese scientist named Ewiti Kurosawa was studying a disease that made rice plants grow abnormally. The rice plants grew very rapidly, becoming much taller than normal, but also producing weak stems that later collapsed and died. The Japanese farmers called these plants *bakane*, which meant foolish. The foolish rice plants were infected with a fungus called *Gibberella fujikuroi*. Kurosawa isolated a chemical from the fungus and then applied it to uninfected plants, resulting in the same rapid growth as seen in the infected plants. Kurosawa called the substance gibberelin after the fungus that made it. Several years later, other scientists purified the gibberelin and determined its chemical structure.

Ethylene

Ethylene is a gas that's best known for its ability to trigger ripening in fruit. People have suspected the existence of ethylene since ancient times: The ancient Egyptians gassed figs to speed up their ripening, and the Chinese burned incense in rooms with ripening pears. Today, commercial growers of many fruits, such as bananas, mangoes, and melons, harvest the fruits when they're green so that they'll be firm for shipping and then gas them with ethylene to give them a ripe color before selling them.

Ethylene also plays an important role in other growth responses, such as triggering leaf abscission, which is the release of leaves from trees.

Brassinosteroids

Brassinosteroids are steroid hormones that control many aspects of plant development. Scientists have found these hormones in a wide variety of plants, including gymnosperms, monocots, and dicots, and even in green algae, which are close relatives to plants. Although brassinosteroids are the most recently discovered group of plant hormones, scientists have worked rapidly to learn about these hormones because of their potential use in agriculture, due to their ability to increase photosynthesis (see Chapter 7) and plant resistance to stress and disease. When brassinosteroids are applied to crop plants, they improve both the yield and quality of the crop.

Brassinosteroids are extremely powerful and can affect the concentrations of other plant hormones. They can either promote or inhibit cell elongation, depending on the stage of plant development and the relative concentrations of auxin and cytokinins. Brassinosteroids may also decrease some of the effects of abscisic acid, acting to promote flowering, seed germination, and opening of stomates.

Which Way Do I Go?: Plant Movements

Plants may not be able to get up and go, but they do move. Plants can change the angle of their leaves so that the leaves track the movement of the sun in the sky. Plants can bend their stems toward the light, too, and their roots grow downward in response to gravity. Some plants even have quick movements, like those of the Venus flytrap, which enable them to respond to the presence of predators and prey.

Growth movements (tropisms)

One type of plant movement — a *tropism* — is a directional movement that results from growth in a particular direction. These growth movements occur in response to elongation of selected cells. Plants exhibit tropic growth in response to many signals, including light, gravity, chemicals, and temperature. As examples, three of the best studied tropisms are

- **Phototropism:** If light shines on a plant from one side, the plant shoot will exhibit *positive phototropism,* and bend toward the light. If you've ever grown plants near a window, you've probably seen evidence of this particular tropism. Molecules called *photoreceptors* in the tips of plant shoots absorb blue light. Signal transduction occurs, leading to a greater production of auxin on the dark side of the shoot. Auxin promotes cell elongation, so the cells on the dark side elongate faster than the cells on the light side and the shoot bends toward the light.

- **Geotropism (gravitropism):** Plant roots grow toward the pull of gravity, so they show *positive geotropism.* Plant shoots, on the other hand, grow away from gravity, showing *negative geotropism.* You can test this by laying a plant on its side. After a while, the shoots will turn and grow upward, while the roots will turn and grow downward. A great deal of evidence supports explanation for geotropism outlined by the *starch-statolith hypothesis.* According to this hypothesis, plants sense the pull of gravity on amyloplasts located in the cells of the root cap and close to the vascular bundles in shoots. The amyloplasts get pulled by gravity so that they collect on one side of the cell that contains them. Auxin is produced in response to the pressure of the amyloplasts, and transported to the side of the root or shoot where the amyloplasts have accumulated. In shoots, auxin stimulates cell elongation, so the side of the shoot where the amyloplasts have accumulated grows faster, and the shoot curves up. In roots, high concentrations of auxin actually inhibit cell elongation, so the side of the root with the amyloplasts grows more slowly than the other side, and the root curves downward.

✔ **Thigmotropism:** I like to think of this tropism as "thing-mo-tropism" because it's curvature in response to touching a thing. One of the most familiar examples of thigmotropism is when a climbing plant, such as a pea or a vine, curls a tendril around a support. Because these plants curve toward the solid surface, they exhibit *positive thigmotropism.* Roots, on the other hand, will curve away from solid objects like the rocks that they encounter in the soil, so they exhibit *negative thigmotropism.* The mechanisms for thigmatropism are still being investigated, but some things are known. Some extremely rapid curvature responses result from changes in turgor pressure). Prolonged curvature results from differential growth, where one side of the shoot or root grows more rapidly than the other. The touch signal is received by epidermal cells, and then signal transduction results in differential growth of the shoot or root. Unlike phototropism and gravitropism, thigmatropism does not seem to be controlled by auxin.

Turgor movements

Turgor movements occur in response to changes in turgor pressure in certain cells. Whereas growth movements take some time because the plant actually has to grow, turgor movements are very rapid. For example, some tendrils exhibit rapid coiling around a solid support, positive thigmatropism, due to turgor movements, followed by continued curved growth due to growth movements. (For more details on curved growth, see the earlier "Growth movements" section)

During turgor movements, cells rapidly lose their turgor pressure and collapse, causing the plant to fold around those cells. Here's what scientists think is happening to cause these responses in tendrils:

1. **Touch receptors sense the pressure of a solid object.**

2. **Signal transduction occurs, causing cells near the touch stimulus to export potassium ions (K^+) outside the cells, thus making the environment hypertonic to the plant cells.**

3. **Water leaves the plant cells, following the K^+, and the cells collapse, causing the plant to fold.**

In addition to curving tendrils, turgor movements also occur in *Mimosa pudica,* which is sometimes called the sensitive plant or the shy plant. If you run your fingers along the leaves of this plant, they fold up immediately. At the base of each leaflet is a swollen structure called a *pulvinus.* When you touch the leaves of the sensitive plant, the signal is relayed to the pulvinus. Cells on one side of the pulvinus lose their turgor pressure, causing the leaves to fold together.

What Time Is It?: Sensing the Seasons

Plants know when to flower, when to go dormant, and when to germinate from their seeds. Both temperature and light play a role in the ability of plants to track the changing days and the changing seasons. Many seasonal responses of plants rely upon the activity of a photoreceptor called phytochrome.

Phytochrome is a photoreceptor that detects the presence of two specific kinds of light: red light (660nm) and far-red (730nm) light. (To see how these types of light fit into the electromagnetic spectrum, see Chapter 7.) Red light is available to plants only during the day when the sun is shining, but far-red light is available both day and night.

As phytochrome absorbs light, it changes between two forms: P_{red} (P_r) and $P_{far-red}$ (P_{fr}). You can think of phytochrome as a molecular switch that flips back and forth between its two forms.

Figure 10-2 illustrates how phytochrome changes form:

- When P_r absorbs red light, it's rapidly converted to P_{fr}.
- When P_{fr} absorbs far-red light, it's rapidly converted to P_r.
- If a plant is in the dark, P_{fr} will slowly convert back to P_r.

P_{fr} is the active form of phytochrome. In other words, the presence of P_{fr} in plant cells is what triggers responses. P_r is inactive.

Phytochrome helps regulate a number of plant responses, including flowering, seed germination, *etiolation* (growing tall in response to shade), as well as plastid and pigment production.

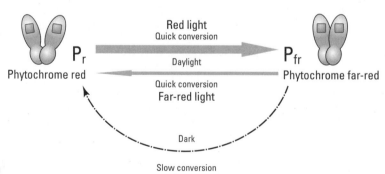

Figure 10-2: Conversion of phytochrome between its two forms: P_{red} and $P_{far-red}$.

Flowering

Many flowering plants use phytochrome to figure out when to make their flowers. Plants use phytochrome to measure the *photoperiod*, which is the length of the light period.

When the sun is shining, phytochrome absorbs the sunlight. As a result, most of the phytochrome in plant cells is converted to the P_{fr} form. But when it's dark, the P_{fr} slowly converts back to P_r. So, the longer the night, the more P_{fr} that gets converted back to P_r. Based on the ratio of P_{fr} to P_r, plants can essentially tell how long the nights are relative to the days.

Plants detect the changing length of the dark period, called the *skotoperiod*, as measured by the conversion of P_{fr} to P_r, in order to measure the time of year.

Botanists divide plants up into three categories based on their flowering preference:

- **Short day plants** flower as the nights get longer and the days get shorter — in other words, in the fall or spring.
- **Long day plants** flower as the nights get shorter and the days get longer — in other words, in the summer.
- **Day neutral plants** flower when they're mature, regardless of photoperiod.

Botanists did experiments to figure out whether it was the length of the photoperiod or the skotoperiod that was important to plant flowering. They used soybean, which is a short-day plant:

- Soybeans that were given 16 hours of light and 8 hours of dark remained vegetative and did not produce flowers.
- Soybeans that were given 8 hours of light and 16 hours of dark produced flowers.
- Soybeans that were given 8 hours of light and 8 hours of dark remained vegetative.
- Soybeans that were given 16 hours of light and 16 hours of dark produced flowers.
- Soybeans that were given 8 hours of light, 8 hours of dark, then 5 minutes of light, and then 8 more hours of dark remained vegetative.

From these experiments, botanists learned that as long as plants experience a continuous dark period that's longer than their unique critical period of darkness, they will flower.

These types of experiments led to the *hourglass model of photoperiodism.* According to this model:

1. **Plants detect the length of the skotoperiod with the phytochrome in their leaves.**
2. **The changing ratios of P_r and P_{fr} signal to the plant whether the nights are getting longer or shorter, thus indicating the season of the year.**
3. **When the plant detects the proper season, the plant leaves relay a signal to the meristems, causing the plant to build reproductive buds (flowers) rather than vegetative buds (leaves).**

The ability to flower can also be affected by the maturity of the plant and by the temperature of the environment. Some plants must grow in a vegetative state for a certain period of time before flowering is possible, regardless of the photoperiod. Other plants must experience *vernalization,* which is exposure to a cold period for a certain length of time, before they'll respond to their correct photoperiod.

Botanists still don't understand all the mechanisms that induce flowering. Recent research on photoperiodism suggests that, in some plants at least, flowering is also affected by the plants' circadian rhythms (see the next section). Botanists have been hunting for the mysterious signal molecule that travels from the leaves to the buds to cause the switch to flower production since 1865 and only recently uncovered its identity! The unknown signal was given the name *florigen* because it causes the beginning (genesis) of flowers. Botanists have identified three genes that are involved in florigen production, called CONSTANS (CO), FLOWERING LOCUS T (FT), and FLOWERING LOCUS D (FD). The FT gene appears to contain the blueprint for florigen, which turns out to be a protein. CO contains the blueprint for a protein called a *transcription factor* that turns on production of the florigen protein. And FD is the blueprint for another protein that works along with florigen in the apical meristem of plants. After it's produced, florigen travels via the phloem from the leaves to the buds, where it triggers the switch from vegetative growth to reproductive growth.

Circadian rhythm

All living things have *endogenous rhythms,* rhythms that persist under constant environmental conditions. When organisms are placed in constant environmental conditions like continuous darkness, their rhythms may become irregular, but when organisms are exposed to natural cycles of light and dark, their rhythms become very regular.

Circadian rhythms are endogenous rhythms that cycle approximately every 24 hours. Some example of circadian rhythms are cycles of melatonin in your body, animal activity, such as that of cockroaches that run at night, and leaf movements of plants that extend their leaves outward during the day and then drop them at night. Part of the reason that people experience jet lag when traveling is that their endogenous circadian rhythm becomes out of sync with the natural light cycle that they've traveled to. It takes a couple of days to re-sync their endogenous rhythm with their environment. Circadian rhythms are set by proteins called *clock proteins* whose concentrations in cells rise and fall at regular intervals.

Recent research suggests that circadian rhythms may play a role in flowering. The mustard plant *Arabidopsis thaliana* is a long day plant that produces flowers in the summer. *A. thaliana* uses phytochrome and another photoreceptor called *cryptochrome,* which responds to blue light, to measure the length of the photoperiod and time its flowering.

Botanists have discovered that these photoreceptors interact with the circadian rhythm of an important control protein called CONSTANS that's needed to trigger flowering.

Seed germination

Seeds are dormant, waiting for the right conditions before they *germinate* or begin to grow.

As seeds develop in the parent plant, seed dormancy is induced by

- The accumulation of abscisic acid, which inhibits growth
- The breaking of connections between the seed and the xylem and phloem of the parent, so that the seed dries out
- The formation of hard, restrictive seed coat
- Increasing levels of acidity in some fruits, such as grapefruit and tomato

To trigger germination, the conditions that promoted dormancy must be reversed:

- A period of cold, wet weather will cause abscisic acid to break down and will increase the levels of gibberellin, which will promote germination.
- Damage to the seed coat from the digestive tract of animals or physical cracking from frost may make it possible for the embryo to begin to grow.

Many botanists use the example of grass seeds like those of barley as an example of the early events that occur during seed germination:

1. **Once the seed is released from dormancy, the embryo releases gibberelin.**
2. **The gibberelin travels to a layer of live cells, called the *aleurone layer*, that wraps around the endosperm in the grain.**

   For the structure of a monocot grain, see Chapter 5.
3. **The gibberelin signals the cells in the aleurone to produce the enzyme alpha-amylase.**
4. **The alpha-amylase breaks down the stored starch in the endosperm, converting it to sugars that the embryo can use to begin growing.**

Some seeds, like those of some varieties of lettuce, require light signals in order to germinate. These seeds detect the presence of light using phytochrome:

- If the seeds are in the dark because they're deep in the soil, the phytochrome will convert from P_{fr} to P_r, and the seeds won't germinate.
- Even if the seeds are in the shade, they may not germinate because the leaves above them will absorb all the red light for photosynthesis, allowing only far-red light to reach the seeds. The far-red light will convert the phytochrome from P_{fr} to P_r, and again the seeds won't germinate.
- If the seeds receive sunlight, then most of their phytychrome will convert from P_r to P_{fr}, and the seeds will germinate.

Botanists tested the reaction of certain varieties of lettuce seeds to light by giving them alternating light flashes of red light and far-red light. It didn't matter how many flashes they gave the seeds; if the last flash was red light then the seeds germinated because red light converts P_r to P_{fr}. If the last flash was far-red light, then the seeds didn't germinate because far-red light converts P_{fr} to P_r. So, if you did a similar experiment and gave your lettuce seeds red light, far-red light, red light, far-red light, red light, far-red light, red light, far-red light, then red light, they'd germinate!

Part III
Making More Plants: Plant Reproduction and Genetics

In this part . . .

If you want to talk about sex, you're in the right part. All joking aside, plants can reproduce asexually by making new individuals that are exact copies of the original plant, and sexually by producing sperm and eggs. When plants reproduce, they pass copies of their genetic information to the next generation. The patterns of inheritance in plants can be predicted by using the laws of genetics first discovered by Gregor Mendel's study of peas. In this part, I outline the basic rules of genetics and talk about the basics of plant reproduction.

Chapter 11

Greening the Earth: Plant Reproduction

In this Chapter
- Comparing asexual and sexual reproduction
- Making more cells with mitosis
- Producing haploid cells with meiosis
- Getting into plant life cycles

Plants are very versatile in that they can reproduce both asexually and sexually. During asexual reproduction, plants use mitosis to make offspring that are genetically identical to themselves. During sexual reproduction, plants use meiosis to produce haploid cells that eventually lead to the formation of gametes. Plant life cycles are complex because they contain two separate alternating generations – one haploid, and one diploid. In this chapter, I explore the pros and cons of different types of reproduction and present the details on exactly how plants do it.

Reproducing: More Than One Way to Do It

Plants are extremely successful organisms in part due to their reproductive flexibility. People can reproduce only by getting together with a mate, but plants have more options. While some plants do reproduce sexually, many plants produce offspring all by themselves by growing underground stems or bulbs that generate new shoots.

Asexual reproduction

If you've ever put a plant cutting in a glass of water to get it to root, then you've seen first hand that some plants can reproduce themselves without any help from another plant. New plants that grow from pieces of old plants are called *clones* because they're genetically identical to their parent. (Imagine if you dropped pieces of yourself and they grew into little "mini-me's"!)

Many plants can do *asexual reproduction,* producing offspring that are identical to themselves (see Figure 11-1). During asexual reproduction, plant cells divide to make exact copies of themselves by a process called *mitosis*.

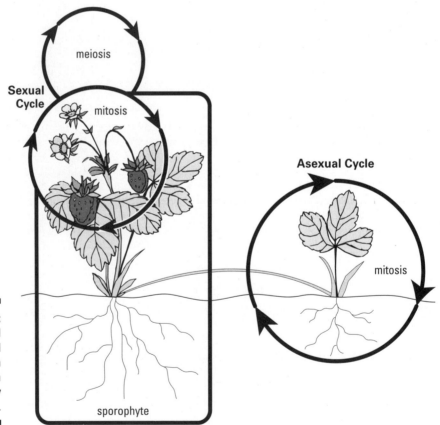

Figure 11-1: Asexual and sexual reproduction in a strawberry plant.

Sticking with a good thing

Corn is a huge important agricultural crop in the United States. If you don't believe me, take a walk through your grocery store and read some labels. Almost everything you can find will contain corn or corn products, such as corn syrup and cornstarch. Even the meat likely comes from animals raised on feed that contains corn! Farmers and agricultural scientists have bred corn for years, trying to develop varieties that have high yield, resist pests, or taste really good. One problem that farmers and agricultural scientists face is that even when they get a variety that they really like, sexual reproduction shuffles up the genes so that offspring may not have all the desired qualities. Enter *agamospermy*.

Some plants, including citrus plants and dandelions, can produce seeds without going through the trouble of sexual reproduction. The seeds are genetically identical to the parent and have all the same traits. Some plant breeders have realized the enormous potential of agamospermy for crop plants like corn — if agamospermic strains could be developed, they would breed true indefinitely.

So, breeders set to work trying to develop agamospermic strains of corn. They started by mating corn with one of its relatives, a grass that can reproduce by agamospermy to create an agamospermic hybrid. This hybrid could reproduce by agamospermy, but it wasn't corn. So, then they did 5,000 matings, each time mating the progeny back to corn to create a plant that was more and more corn-like but could still reproduce by agamospermy. The breeders were successful, up to a point. They did create agamospermic corn, but the yield on the variety is not high enough for commercial purposes. Other breeders and scientists are very interested in this work, however, because it shows what might be possible in the future. Some scientists hope to isolate the genes that are necessary for agamospermy, then figure out how to introduce them directly into plants. This might create agamospermic varieties without having to go through 5,000 matings!

Asexual reproduction occurs by a variety of methods in plants and can involve different organs:

- **Stems:** Stems may extend from the parent plant, forming new individuals at intervals. Strawberries, such as the plant in Figure 11-1, reproduce asexually by stolons, while potatoes use tubers, daffodils grow from bulbs, and gladioli reproduce by corms. (For the details on these modified stems, see Chapter 4.) In the case of potatoes, the eyes of the potato are actually buds that can grow into new potato stems. The bulbs of daffodil and gladiolus plants get bigger year after year, forming lateral buds that grow into new stems.

- **Roots:** Some plants, such as aspen and poplar trees, send roots out horizontally from the original plant. The horizontal roots grow new stems at intervals, producing large stands of trees that are genetically identical to each other. Other plants, such as dandelion, have a long tap root that can generate new stems if the original stem is damaged.

- **Leaves:** Leaves of the common houseplant *Kalanchoë* form tiny little *plantlets* along the edges. These plantlets can drop to the ground, produce roots, and grow into new individuals. Plants called liverworts make tiny little cups called *gemmae* on the surfaces of their leaf-like bodies. When rain splashes into the cups, clumps of cells break off and scatter, forming new plants wherever they land.
- **Flowers:** Although flowers usually mean sexual reproduction, some plants, such as citrus trees, produce seeds asexually by a process called *agamospermys*. During agamospermy, a cell from the parent plant develops into a seed containing an embryo that is a genetic copy of the parent.

Fragmentation, the process where a plant part breaks off and develops into a new individual, occurs with stems, roots, and even leaves.

Sexual reproduction

Although asexual reproduction is very common in plants, most plants can also mix it up a little by combining their genetic information with another plant to produce offspring that have unique combinations of genetic information.

Most plants can do asexual reproduction and *sexual reproduction*, where an egg from one plant combines with sperm from another plant to produce offspring that have new combinations of traits. Tulips, for example, reproduce asexually through tulip bulbs, but also reproduce sexually by making flowers, fruits, and seeds.

The most obvious participants in sexual reproduction are the flowering plants — their showy flowers are basically advertising that they're looking for a mate — but more subtle plant sex is happening around you all the time. All the cone-bearing plants, such as pine, spruce, and fir, do sexual reproduction, as do flowering trees like maple, oak, and willow. Sex is even going on among the grass and moss under your feet!

Comparing reproductive styles

Both asexual reproduction and sexual reproduction have their advantages:

- **Asexual reproduction rapidly produces copies of a successful organism.**
  - **Advantage:** If the traits of the parent plant are working well in a particular environment, then asexual reproduction will produce more successful organisms without the hassle of finding a mate. In other words, if it ain't broke, don't fix it.
  - **Disadvantage:** If conditions change so that the parent's traits aren't very successful, then all the clones will be negatively impacted at once. Also, small genetic changes called *mutations* that occur in the clones will be passed onto all their offspring. Most mutations have negative impacts on individuals, asexually reproducing populations collect more and more mutations over time.
- **Sexual reproduction creates individuals with unique combinations of traits, increasing the chances that some offspring will survive.**
  - **Advantage:** Environments change, and individuals come under attack from predators and parasites. The greater the diversity in offspring, the more likely it is that some will have traits that help them survive. And mutations that occur in parents may not get passed on to offspring at all or may get passed to just some offspring, so that some offspring will be free from these mutations.
  - **Disadvantage:** Offspring from sexual reproduction may be more fragile at first than offspring from asexual reproduction. (Think tiny seedling versus new shoot supported by an underground stem.) And, of course, you can't go it alone — sexual reproduction means successfully capturing gametes from another individual.

Copying Cells by Mitosis

Eukaryotic cells divide by mitosis in order to make exact copies of themselves. Plants use mitosis in order to grow, repair tissues and reproduce asexually.

Mitosis occurs at different rates in different plant tissues. Cells that are part of meristems divide frequently, whereas mature cells in the vascular tissue never divide (in fact, some die at maturity). Other cells may not typically divide, but might divide to repair damage. So, plant cells may spend some time dividing, and some time not dividing.

The dividing phase of eukaryotic cells is called *mitosis,* and the nondividing phase is called *interphase.* The alternating cycle of mitosis and interphase is called the *cell cycle* (see Figure 11-2).

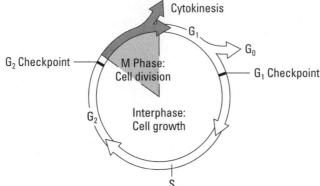

Figure 11-2: The cell cycle.

Interphase

Interphase contains three subphases:

- **Gap 1 (G1):** Plant cells that are alive and functioning, but not dividing, are in the phase called *Gap 1 (G_1)*, which is the phase that cells spend most of their time in. A leaf mesophyll cell, for example, will hang out in G_1 doing photosynthesis. If a cell gets a signal that they're going to divide, they'll grow and double their contents.

   Cells that are going to divide must pass a test, called a *checkpoint*, before they can exit G1 and enter the next phase of interphase.

   Checkpoints are points in the cell cycle where cells check to make sure that everything is proceeding normally. If cells can't pass a checkpoint, repairs will be made, if possible. If not, the cell may be signaled to commit cell suicide, called *apoptosis*. In order for plant cells to pass the G1 checkpoint, they must be big enough, have plenty of nutrients, and have undamaged DNA.

- **Synthesis phase (S phase):** S phase is called the synthesis phase because cells synthesize, or make, new DNA. During S phase of Interphase, plant cells copy all their chromosomes so that they'll have a complete set to give to a new cell during cell division. Cells attach the two copies of each chromosome together with proteins that stick to a region of the chromosome called the *centromere*.

   After S phase, scientists call chromosomes *replicated chromosomes* because each chromosome is actually made of two identical DNA molecules attached together at the centromere. People usually draw replicated chromosomes as little X's like the ones in Figure 11-3. Scientists call each half of the X a *sister chromatid*. The two sister chromatids of a replicated chromosome join together at the centromere. So, before S phase and after mitosis, chromosomes are single DNA molecules, but between S phase and mitosis, replicated chromosomes are double DNA molecules.

- **Gap 2 (G2):** During Gap 2, cells check the work they did during S phase. In order to enter cell division, cells must pass the G_2 checkpoint. The G_2 checkpoint tests whether the cell copied all the chromosomes correctly. If cells can't pass the G_2 checkpoint, they remain in G_2 and try to repair themselves. If they can't repair themselves, they may kill themselves via apoptosis.

Overview of mitosis

Like all eukaryotes, plant cells have multiple *chromosomes,* individual DNA molecules that store their genetic information. Pea plants, for example, have 14 separate chromosomes per cell, while some wheat plants have 42 chromosomes. When a plant cell divides by mitosis, it has to make sure that each new cell gets a complete set of chromosomes. That's what mitosis is all about: During S phase, plant cells make copies of their chromosomes, so during mitosis, they need to sort carefully and make sure that each cell gets one of each chromosome.

Mitosis is a carefully controlled process that organizes and separates the chromosomes correctly. After the cells separate the chromosomes and build new nuclei, they divide their cytoplasm by *cytokinesis* to form two distinct cells.

The cell structure that separates the pairs of chromosomes is the *mitotic spindle.* The mitotic spindle is made of microtubules, which are cytoskeletal proteins. *Spindle fibers* of microtubules attach to the chromosomes in order to push and pull them during mitosis.

The four phases of mitosis (refer to Figure 11-3) have unique events that contribute to the careful sorting of the chromosomes:

- **Prophase:** During Interphase, the DNA is uncoiled, and the chromosomes aren't visible, even under a microscope. During prophase, cells tightly coil their chromosomes, making the chromosomes easier to move around. Coiling of the chromosomes, called *condensation,* also makes the chromosomes visible under a microscope.

  During prophase, cells break down their nuclear membrane so that the mitotic spindle can reach into the center of the cell and attach to the chromosomes. Cells also break down their *nucleoli,* the dark staining regions in the nucleus that contain the materials for ribosome synthesis. Plant cells can have multiple nucleoli per nucleus. Some scientists separate the events from the breaking down of the nuclear membrane up to the lining up of the chromosomes into a separate phase, called *prometaphase.*

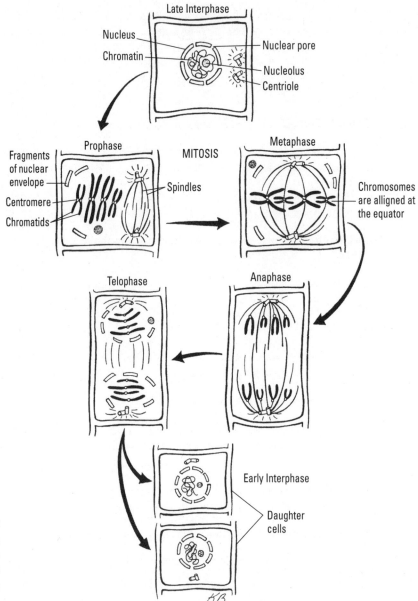

Figure 11-3: The process of mitosis and interphase.

✓ **Metaphase:** During metaphase, cells organize the chromosomes by lining them up in the middle of the cell, halfway between the two poles of the mitotic spindle. At the *metaphase checkpoint,* cells make sure that all the replicated chromosomes are attached to the mitotic spindle. Cells can't continue through mitosis unless they pass the metaphase checkpoint.

✔ **Anaphase:** During anaphase, cells separate the replicated chromosomes so that each sister chromatid goes to opposite sides of the cell. Cells separate the identical sister chromatids carefully to ensure that each new cell will get one. Once the sister chromatids separate from each other, scientists call them chromosomes again.

✔ **Telophase:** Telophase wraps up mitosis by reversing the events of prophase. During telophase, cells uncoil their chromosomes, reform their nuclear membranes and nucleoli, and break down the mitotic spindle.

Cytokinesis

After a plant cell separates the sister chromatids and builds new nuclear membranes to create two nuclei, it divides its cytoplasm into two parts by forming new plasma membrane and cell wall down the middle of the cell. Figure 11-4 shows how cytokinesis works in a plant cell:

✔ The cytoskeleton moves small vesicles containing cell wall material (pectin and hemicelluloses) into a line in the middle of the cell.

✔ The vesicles fuse together, bringing together the vesicle membranes and the wall material that was inside the vesicles. The vesicle membranes form the new plasma membranes for each cell. The wall material joins together to form the *cell plate*. The two new cells then secrete cellulose and other materials to build a primary cell wall on either side of the cell plate, which is now called the middle lamella. (For details on cell wall structure, go to Chapter 2.)

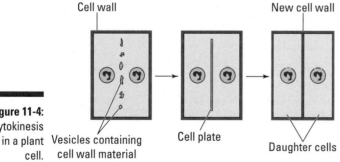

Figure 11-4: Cytokinesis in a plant cell.

Reproducing Sexually with Meiosis

Like all sexually reproducing organisms, plant cells divide by meiosis to produce cells that have half the genetic information of the parent.

Counting chromosomes

Each organism has a unique set of chromosomes. For example, pea plants that have 14 chromosomes per cell actually have two each of seven different types of chromosomes (2 x 7 = 14). For peas, one complete set of chromosomes is a set that contains one each of the seven different types.

Scientists call one complete set of chromosomes the *haploid number* of chromosomes, represented by the letter *N*. So, the haploid number of chromosomes in peas is seven chromosomes (N=7). Each of the seven chromosomes contains different types of genetic information, and each is essential for normal development of a pea plant. Pea gametes are haploid because they have seven chromosomes, or one of each kind, per cell. A zygote receives two of each kind of chromosome, one from each parent. Each matching pair of chromosomes contains the same kinds of genes.

Scientists call matching pairs of chromosomes from the organism's two parents, containing the same kinds of genes, *homologous chromosomes.*

When two haploid pea gametes (7 chromosomes each) fertilize each other, they create a zygote with 14 chromosomes, two each of the seven different kinds (2N=14 — or in other words, seven pairs of homologous chromosomes)

Cells with two complete sets of chromosomes are *diploid.* Scientists refer to diploid cells as 2N.

Diploid cells divide by meiosis to produce haploid cells. Diploid cells can't just pile up all their chromosomes and randomly split them in half, because then the chromosomes wouldn't be sorted into sets. A haploid cell has to have a complete set in order to be normal. So, the whole point of meiosis is to carefully sort the sets of chromosomes to make sure haploid cells are truly haploid — that is, they have one complete set of chromosomes.

Scientists call the number of chromosome sets in a cell has its *ploidy.* Cells in certain plant tissues can have three or four sets of chromosomes, making them triploid, tetraploid, and so on. Some plants are *polyploid,* which means they have many sets of chromosomes per cell. One polyploid fern, *Ophioglossum*, has 1,200 chromosomes per cell!

In plants, the products of meiosis are spores, not gametes that are produced by meiosis in animals.

Spores grow into multicellular haploid gametophytes, which then make haploid gametes by mitosis. That way, when sperm and egg join to form a zygote, the new sporophyte plant that develops will have the normal diploid number of chromosomes for that species.

Following the plan

Meiosis occurs as part of the cell cycle (refer to Figure 11-2), just like mitosis. In fact, meiosis has many similarities to mitosis:

- Cells that are going to undergo meiosis receive a signal that causes them to leave G_1 of interphase and enter S phase.
- During S phase, the cells copy all the chromosomes, creating replicated chromosomes that have identical sister chromatids attached at the centromere.
- The cells then proceed through G_2 and into meiosis.
- During meiosis, the mitotic spindle moves the chromosomes and sorts them in similar ways as it does during mitosis. Scientists even use the same terms — prophase, metaphase, anaphase, and telophase — to describe the movements of the chromosomes.

Doubling your stuff and then reducing it twice

Cells divide by meiosis to cut the amount of DNA in the cell by half to make haploid cells. However, before meiosis occurs, cells double their DNA during S phase. So, in order to get the DNA down to the haploid amount, meiosis includes two divisions. (If you start with double DNA and you want to get to half, you have to divide twice — once to get from double DNA to single DNA, then again to get to half.) Figure 11-5 shows the two divisions of meiosis, called *meiosis I* and *meiosis II*, occurring as part of normal meiosis.

The details of meiosis may seem less overwhelming if you remember the purpose of each division:

- Meiosis I separates the pairs of homologous chromosomes.
- Meiosis II separates sister chromatids.

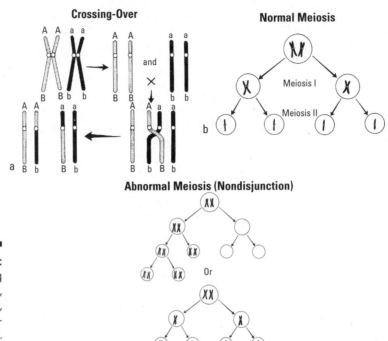

Figure 11-5: Crossing over, meiosis, and nondisjunction.

The events of meiosis I

The phases of meiosis I have many similarities to the phases of mitosis. Figure 11-6 shows the major events of each phase:

- **Prophase I:** Prophase I includes the same major events of prophase of mitosis. Cells in prophase I coil up their chromosomes and break down their nuclear membrane and nucleoli. During prophase I, homologous chromosomes find each other and pair up. The two replicated chromosomes of each pair actually stick together, forming a structure called a *tetrad*. Tetrads have four arms because each replicated chromosome has two sister chromatids.

    While homologous chromosomes are stuck together in tetrads, they can exchange small pieces of DNA in a process called *crossing-over*. Figure 11-5 shows how crossing over leads to new combinations of genetic material in the chromosomes. Because crossing-over creates new combinations of genetic material, it increases the diversity in offspring. This increase in diversity is one of the advantages of sexual reproduction.

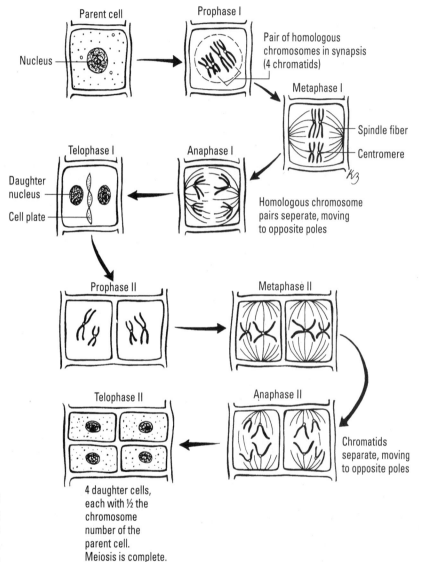

Figure 11-6: The events of meiosis.

- **Metaphase I:** The spindle pushes the homologous pairs of chromosomes to the middle of the cell. The major difference between metaphase I of meiosis and metaphase of mitosis is that chromosomes line up in pairs during metaphase I, and they line up as individuals in metaphase of mitosis.

- **Anaphase I:** Spindle fibers shorten towards opposite sides of the cell, pulling the pairs of homologous chromosomes apart so that one chromosome from each pair goes to opposite sides of the cell.

- **Telophase I:** Cells reform nuclear membranes creating two nuclei and break down the spindle. After telophase I, the two nuclei are haploid because they only have one of each type of chromosome. Cells divide their cytoplasm by cytokinesis, resulting in the formation of two cells.

After meiosis I is complete, both cells proceed directly to meiosis II without going through the stages of interphase.

The events of meiosis II

Meiosis II separates the sister chromatids of each replicated chromosome and sends them to opposite sides of the cell. Once again, the cells pass through several phases similar to those of mitosis and meiosis I, as shown in Figure 11-6:

- **Prophase II:** Cells reform the mitotic spindle and attach it to the chromosomes. If nuclear membranes formed during telophase I, they break down again.
- **Metaphase II:** Spindle fibers push the replicated chromosomes to the center of the cell where they form a line. *Tip:* The major difference between metaphase II of meiosis and metaphase of mitosis is that the number of chromosomes in metaphase II is only half that of metaphase of mitosis.
- **Anaphase II:** Spindle fibers shorten, pulling sister chromatids apart so they move to opposite sides of the cell.
- **Telophase II:** Cells complete formation of the haploid cells by reforming the nuclear membranes and nucleoli and uncoiling their chromosomes. Cytokinesis occurs in both cells, forming a total of four haploid cells.

Considering Alternation of Generations

Plant life cycles are considerably more complicated than animal life cycles. In animals, meiosis produces haploid cells called gametes that don't last very long on their own. Animal gametes fuse with other gametes, returning the chromosome number back to diploid.

It can be really hard for people to relate to plant life cycles because plants do something humans don't do — they produce two different generations within one life cycle, whereas people and all other animals produce one.

To help you relate to the plant life cycle, here's a comparison between people and plants:

✔ **People:** An individual person is like a plant sporophyte because they have a double copy of the genetic information. When people reproduce sexually, cells in their body undergo meiosis to produce sperm or egg. Sperm and egg get together and a new individual is produced. So, if scientists talked about people in plant terms, people would go from sporophyte (parent) to gametes (egg and sperm) and right back to sporophyte (offspring).

✔ **Plants:** Plant sporophytes produce reproductive cells through meiosis, but they don't go straight to egg and sperm. Instead, the cells produced by meiosis grow for a while by mitosis, making a unique plant structure (the gameteophyte), which eventually gets around to making egg and sperm. If scientists talked about plants in people terms, it would be like the parents made an egg- or sperm-like cell that grew to have a little life of its own — maybe even as a separate individual — and then that little gametophyte makes the sperm and egg, which join together to make the new individual (sporophyte).

Figure 11-7 shows how plant life cycles actually have two separate generations:

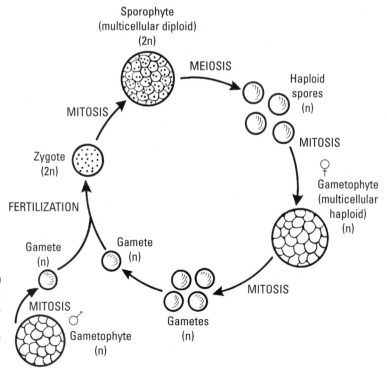

Figure 11-7: Overview of alternation of generations.

- **Gametophyte generation:** Meiosis produces haploid cells called spores (also called *meiospores*). The first cell of the gametophyte generation is the spore. The spores then divide by mitosis to produce a multicellular haploid structure called a *gametophyte* (*gameto* for gametes and *phyte* for plant, so literally a gamete-bearing plant). Cells of the gametophyte generation are haploid. The gametophytes produce gametes by mitosis. Gametes are part of the gametophyte generation, but once they fuse together during fertilization, the sporophyte generation begins.

- **Sporophyte generation:** When gametes fuse together, they create a diploid zygote. The first cell of the sporophyte generation is a zygote. The zygote then divides by mitosis to produce a multicellular diploid structure called a *sporophyte* (*sporo* for spores and *phyte* for plant, so literally a spore-bearing plant). Cells of the sporophyte generation are diploid. Sporophytes produce spores by meiosis, which starts a new life cycle with the gametophyte generation. Scientists call the cell of the sporophyte generation that undergoes meiosis a *spore mother cell* or *meiocyte*.

Because a complete plant life cycle includes both the gametophyte and the sporophyte generations, scientists refer to the plant life cycle as having an *alternation of generations*. Meiosis causes the change from the sporophyte to the gametophyte generation, while fertilization causes the change from the gametophyte to the sporophyte generation.

The balance between the gametophyte and sporophyte generation shifts between different groups of plants. In many of the oldest plant groups, like mosses and liverworts, the gametophyte generation is large and visible. The trend through the evolutionary history of plants, however, is for the gametophyte generation to get smaller and more protected by the sporophyte.

Figure 11-8 shows an example of the life cycle of *Marchantia,* a member of an ancient plant group called liverworts (Chapter 15 presents this group of plants in detail.) In liverworts, the gametophyte generation is dominant. In other words, if you were in a wet place in the woods, maybe near a stream, and saw some of these flat leafy plants growing on the rocks, what you'd be looking at is haploid gametophyte tissue. The sporophyte generation in liverworts is very small, forming in pockets of tissue on the gametophyte and never living independently.

In ferns, which evolved more recently and are vascular plants, the gametophyte generation is still visible, but the sporophyte generation is dominant. When you see ferns growing in the woods or by the roadside, you're looking at diploid sporophyte tissue. To see a fern gametophyte, you'd have to get down on your hands and knees and look around on the ground under the fern fronds. Fern gametophytes aren't very large — about the size of your thumbnail — and look like little green hearts. Figure 11-9 shows alternation of generations in a fern life cycle. (Chapter 15 goes into ferns in more detail.)

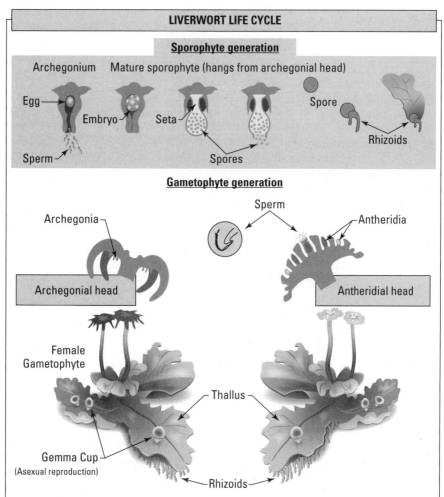

Figure 11-8: Alternation of generations with gametophyte dominant (liverwort as example).

More recent still, in terms of evolutionary age, are the flowering plants. Flowering plants are much more comparable to people in the way they run their life cycle — the main plant is diploid, and the gametophytes are restricted to just a few cells that produce gametes. (Chapter 17 gives you the full scoop on flowering plants).

In order to understand why gametophytes are so tiny in vascular plants, you have to know a little bit about where plants came from. Life began in the ocean, so the earliest plants were surrounded by water. When some intrepid plants started to creep out onto the land, they had to make a lot of adjustments to get the water they needed and keep from drying out. (See Chapter 13 for some of the amazing ways that plants adapt to different environments.) But the way plants made life on land work for them was to make changes in their sporophytes, such developing protective cuticles to retain water and vascular tissue to move water throughout the plants. Instead of changing the gametophytes, plants evolved mechanisms for protecting them. In the earliest vascular plants, such as ferns, gametophytes are small and short-lived, surviving only in the wet season of the year. In seed plants like gymnosperms (see Chapter 16) and flowering plants (see Chapter 17), the gametophytes can survive in dry conditions because they're totally enclosed in sporophyte tissue.

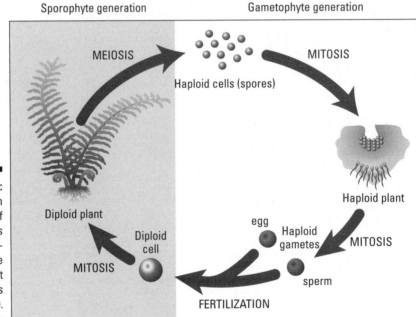

Figure 11-9: Alternation of generations with sporophyte dominant (fern as example).

Chapter 12

Passing Plant Characteristics to the Next Generation

In This Chapter
▶ Discovering the laws of inheritance with Gregor Mendel
▶ Digging into monohybrid and dihybrid crosses
▶ Mastering the lingo of genetics

Through genetics, scientists and doctors can understand the transmission of genetic diseases, farmers can breed the best crop plants and animals, and everyone can appreciate how their own traits connect to those of their families. Everything scientists know today about genetics began with plants and the breeding experiments of an Austrian monk named Gregor Mendel. This chapter presents the fundamentals of genetics that were originally revealed by Mendel and looks at some of the unique features of plant genetics.

Tracking the Inheritance of a Single Gene

Genetics is the study of inheritance — in other words, tracking traits from one generation to the next. Plant cells contain pieces of DNA, called *chromosomes*, that determine the traits of the organism. A *gene* is a section of a chromosome that contains the code for a worker molecule, such as RNA or proteins, in a cell. (For more on DNA, RNA, and proteins, see Chapter 2.)

Each chromosome within a cell may contain hundreds of individual genes. Cells use their genes as blueprints to build the molecules that determine their structure and function. So, an organism's genes lead to certain molecules that define the traits of the organism. For example, a gene in a flowering plant might contain the information for building an enzyme whose job it is to produce a colored pigment in the flower of the plant. Depending on the code in the genes of a plant, the plant may have purple or white flowers.

You can think of each set of chromosomes in a cell as a file cabinet. The drawers in the file cabinets represent the chromosomes, and the files in the drawer represent the genes contained on that chromosome. When a cell needs to build a particular worker molecule, it goes to a particular drawer (chromosome) and finds the file (gene) that contains the blueprint for that molecule. The cell then uses the blueprint as instructions for building the molecule.

Most plant cells are diploid, meaning they have two full sets of chromosomes, so they have two copies of every gene. You can think of a diploid cell as having two complete file cabinets, one copied from Mom's files and one copied from Dad's. The file cabinets contain matching drawers, or chromosomes. In other words, for every drawer in the file cabinet from Mom, a matching drawer in the file cabinet from Dad contains the same kinds of information. If the drawer from Mom contains files (genes) for making molecules that determine flower color, then the drawer from Dad also contains files that determine flower color.

The combination of genes an organism receives from its parents interacts with each other to determine the unique characteristics of the offspring. For example, if a flowering plant receives genes that specify purple flowers from both Mom and dad, then the offspring will have purple flowers. If a flowering plant receives genes that specify white flowers from both Mom and Dad, the offspring will have white flowers. These examples, where the offspring look just like the parents, probably seem pretty simple to you. But what if one parent gives a message for purple flowers and the other parent gives a message for white flowers? To understand the answer to that question, you need to take a look at the work of Gregor Mendel, who did exactly that experiment with purple- and white-flowered pea plants.

Investigating plants with Gregor Mendel

Gregor Mendel was an Augustinian monk who lived in the 1800s, way before anyone knew anything about chromosomes, DNA, genes, or even meiosis. Yet, without knowing anything about the cellular components involved, Mendel was able to figure out the fundamental rules that govern inheritance. Mendel made his great discoveries by spending seven years carefully observing, counting, and analyzing the results of breeding experiments in pea plants. Although many people had done breeding experiments with crop plants and animals before Mendel, one thing that made his work unique was that he carefully applied mathematical analysis to his work. The ideas that Mendel proposed to explain the mathematical patterns he saw in his breeding experiments have withstood the test of time and become the founding principles of the science of genetics.

Mendel did his work on pea plants in the garden of the abbey where he was a monk. He studied many different traits of pea plants, including flower color, pea shape, pea color, pod shape, and plant height. Peas were very useful for

Chapter 12: Passing Plant Characteristics to the Next Generation

his experiments because pea plants are *self-fertile:* Pea flowers contain both male and female parts and can make offspring after being mated to themselves. If Mendel wanted to ensure that a pea plant didn't reproduce with itself, he snipped the male parts off the flowers.

So, Mendel was able to *outcross* pea plants by mating pea plants to other pea plants and also *self-cross* pea plants by mating them to themselves:

- To do a self-cross, Mendel brushed pollen from a flower onto the female parts of the same flower.
- To do an outcross, Mendel brushed pollen from one plant onto the female parts of a different plant.

After performing a mating, Mendel waited for the plants to produce pea pods (the fruit), containing peas (the seeds). Then he collected the offspring peas and planted them. After the offspring grew, he recorded their traits.

Before he began his experiments, Mendel used the ability of pea plants to self-cross to create *pure-breeding* lines of plants, plants that always produced the same version of a trait in the offspring between two pure-breeding individuals.

For example, Mendel created pea plants that were pure-breeding for flower color — purple-flowered plants that always produced purple-flowered offspring, and white-flowered plants that always produced white-flowering offspring. He made the pure-breeding plants by self-crossing plants of each color for many generations, always removing any offspring that didn't have the right flower color, until the plants consistently produced only one color flower. By creating pure-breeding plants, Mendel knew he was starting any experiment with plants that gave a certain message to their offspring.

To begin one of his experiments, Mendel chose pure-breeding parents that had different variations of a trait he was interested in and then followed the trait for several generations. To keep track of the generations in a genetic cross, such as the one Mendel performed between purple- and white-flowered peas (see Figure 12-1), scientists use a specific notation:

- **The parental generation is identified as P_1.** P_1 plants are always pure-breeding for the trait that is being studied. For example, to study the inheritance of flower color, Mendel crossed pure-breeding purple-flowered peas with pure-breeding white-flowered peas.

- **The cross between the parentals produces the first generation of offspring, called the F_1 generation.** F stands for filial, which refers to children, so the F_1 generation is the *first filial generation* or the first generation of offspring from the parental cross. Mendel collected the peas produced by his parental cross, planted them, and observed the traits of the offspring. In the case of purple flower crossed with white flower, all the F_1 plants made purple flowers.

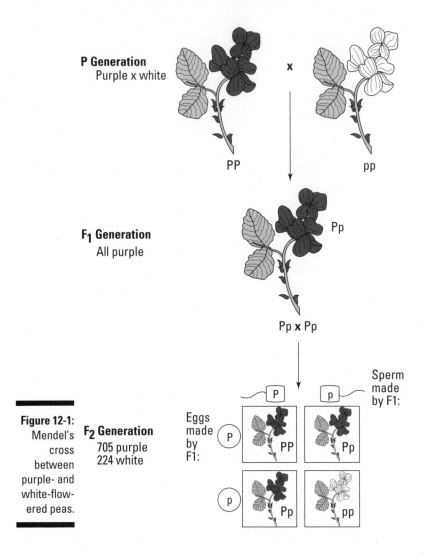

Figure 12-1: Mendel's cross between purple- and white-flowered peas.

- A cross between two members of the F_1 generation produces the second generation of offspring, called the F_2 generation. F_2 stands for *second filial generation*. Mendel crossed different individuals in the F_1 generation with each other, collected the seeds, planted them, and observed the traits of the F_2 generation, which produced 705 purple-flowered plants and 224 white-flowered plants.

Figuring out the rules of inheritance

Mendel noticed several important things about his crosses that led him to propose some very big ideas that scientists now call *Laws of Inheritance:*

Chapter 12: Passing Plant Characteristics to the Next Generation

- The *Law of Unit Characters,* **which is also referred to as the idea of** *particulate inheritance,* **states that each plant has two copies of every gene and that each pair of genes controls the inheritance of various characteristics.** Scientists call the different variations of genes *alleles.* For example, for the gene that controls flower color in peas, each plant may have two alleles for purple color, or it may have two alleles for white color, or it may have one allele for each color.

    - **What Mendel saw:** A variation of a trait, such as white flowers, could disappear in the F_1 generation and reappear in the F_2 generation.

    - **What Mendel concluded:** Each plant must have two factors that control each trait. Even though all the plants in the F_1 generation appear to be the same and only show the effect of one factor, the other factor appears again in the F_2 so it must be present in the F_1.

- **The *Law of Dominance* says that for any pair of alleles, one allele may mask the effect of another.** Scientists describe the observable allele as *dominant* and the hidden allele as *recessive*.

    - **What Mendel saw:** One variation of a trait may hide another variation. For example, plants of the F_1 generation shown in Figure 12-1 must have received a message for purple flowers from one parent and a message for white flowers from the other parent, but they all show purple flowers.

    - **What Mendel concluded:** For each pair of factors that control a trait, one factor may hide the effect of another.

- **The *Law of Segregation* says that each pair of alleles separates during meiosis, so that gametes receive one of each pair.** The *somatic,* or body, cells of a parent are diploid, but the gametes produced by a haploid gametophyte are haploid.

    - **What Mendel saw:** In the F2 generation, the ratio of plants with the dominant trait to plants with the recessive trait was approximately 3:1. For example, the cross shown in Figure 12-1 produced 705 purple-flowered plants and 224 white-flowered plants, which you can write as a ratio of 705:224. If you divide both numbers by 224 to reduce the ratio, you get a final ratio of 3.15:1. This 3:1 ratio is what you would expect if F_1 parents gave one copy of each message to their F_2 offspring. (See the squares at the bottom of Figure 12-1.)

    - **What Mendel concluded:** Parents have a pair of factors that control each trait, but they pass only one copy of each factor to their offspring.

Segregation of alleles occurs during meiosis I, when homologous chromosomes separate from each other. (See Chapter 12 for more on meiosis.)

Many people have a common — but wrong — idea that dominant alleles are more common in populations. On the surface, this idea seems reasonable — if an allele can hide another allele, it will show up more, right? If the two alleles were present in equal amounts in the population, that assumption would be true, but it isn't necessarily the case. The only way to know whether a trait is dominant or recessive is to look at an individual that has two different alleles for a single trait. The appearance of the trait in that individual tells you whether the trait is dominant or recessive.

Speaking like a geneticist

Since Mendel's day, scientists have learned a lot more about inheritance, and they've developed a whole language and standard symbols of genetics to describe their thinking. In order to understand what scientists are saying and solve problems in genetics, you need to master some fundamental terms:

- **Chromosome, gene, allele, and locus:** I already defined chromosome, gene, and allele in the previous sections, but it's probably still helpful to you if I put them in context relative to one another. Chromosomes contain coding regions (or DNA sequences) called genes. Variations in the code of a gene lead to different alleles of the gene. The place where a particular gene is located on a chromosome is called a *locus*.

  - *Plant geneticists represent different alleles by using letters of the* alphabet — usually the first letter of the trait being studied. For example, a plant geneticist might use the letter P to represent the purple allele for flower color in peas.

  - If a gene has two alleles, plant geneticists use the same letter for both alleles of the trait, but they indicate the dominant allele with a capital letter and the recessive allele with a lowercase letter. So, if the purple allele is P, then the white allele is p (and not w).

- **Wild-type and mutant:** Wild-type is the normal, or typical, form of a trait, found in the majority of individuals of a species in nature. Mutant is anything different from wild-type.

- **Genotype and phenotype:** The alleles an organism has for a particular gene is called its *genotype*. The way the organism appears and functions because of its alleles is its *phenotype*. The genotype is the blueprint for the organism; the phenotype is what is built from that genotype. In Figure 12-1, the genotype for flower color of the F_1 generation is shown as Pp, indicating the alleles for the trait. The phenotype of the F_1 generation is how the plants look for this trait, which is "purple flowers."

- **Homozygous and heterozygous:** Organisms that are homozygous for a gene have two identical alleles. Organisms that are heterozygous have two different alleles. In Figure 12-1, the parentals are both homozygous for flower color (written PP or pp), while the F_1 generation is heterozygous (Pp). Organisms with two different alleles are also commonly called *hybrids*. An organism that is heterozygous for one gene is called a *monohybrid*.

The crosses between individuals in the F1 generation are called *monohybrid crosses* because they are matings between two individuals that are heterozygous, or hybrid, for one trait. Homo means same, so homozygous is having the same two messages for a gene. Hetero means different, so heterozygous is having two different messages for a gene.

Making predictions

The square shown at the bottom of Figure 12-1 is a *Punnett square,* a tool that geneticists use to diagram and predict the results of a genetic cross. On the edges of a Punnett square, you write symbols for the alleles in the gametes that are combining in the cross. By filling in the square, you create all the possible combinations of gametes that may occur in the next generation.

For example, the Punnett square in Figure 12-1 shows the predicted outcome of a monohybrid cross. Each member of the F_1 generation is heterozygous, having two different alleles for flower color (Pp). When these plants make eggs and sperm, they will give one allele to each gamete (either P or p).

To make the Punnett square for a monohybrid cross:

1. **Use the parents' genotypes to figure out the genotypes for the gametes they produce.**

    Each gamete receives only one allele for each pair off alleles in the diploid parent. A diploid parent that has the genotype Pp will ultimately produce gametes that have either the genotype P or the genotype p.

    In plants, diploid sporophyte parents don't directly make gametes — first they make haploid spores by meiosis that go on to make gametes by mitosis. Because the gametophytes and the gametes have the same genotype, it's easier to just think about the ultimately products of meiosis being gametes.

2. **Write the genotypes for the gametes from each parent along one side of the Punnett square.**

    Eggs, which can either be P or p, are written along one side of the square; sperm, which can also either be P or p, are written along the other side of the square.

A wonder of a weed

Gregor Mendel worked with peas, but most of today's plant geneticists work with a tiny little mustard plant that could be growing like a weed right now in a field near you. *Arabidopsis thaliana*, sometimes called the mouse-ear cress, has become the lab rat of plant genetics. *A. thaliana* is great for studies in genetics for lots of reasons: It's tiny, so it can be grown easily in labs and greenhouses, it completes its life cycle from seed to seed in about 6 weeks so scientists don't have to wait long to get the results of a cross, and it produces lots and lots of offspring, as many as 15,000 seeds per plant. And, *A. thaliana* might look like a little weed to you, but to scientists, it still qualifies as a flowering plant. And importantly, it has many of the same genes as other flowering plants, including those that have agricultural interest to people.

A. thaliana is also easy to work with because, as far as plants go, *A. thaliana* doesn't have that many genes, enabling scientists to figure out the sequence of its entire genome — in other words, to read the code of all its 115,409,949 base pairs of DNA and determine that it has a total of about 28,000 genes. By studying the genes of *A. thaliana*, scientists learn about the genes of plants that are valuable to people. Through studies of this little weed, scientists have learned a great deal about the genes that control plant growth, seed production, pest resistance, flowering — you name it, really. Organisms like *A. thaliana* that are studied in labs as a way to know more about life in general are called *model organisms*. And *A. thaliana* is certainly a model that good things can come in small packages.

3. **Write the combinations of alleles that result from each type of egg combining with each type of sperm to create the predicted genotypes for the F_2 generation.**

    If a P egg is fertilized by a P sperm, the result will be an offspring that is PP. If a P egg is fertilized by a p sperm, or if a p egg is fertilized with a P sperm, the result will be offspring that are Pp. If a p egg is fertilized by a p sperm, the offspring will be pp.

  The ratio of genotypes predicted in the F2 is 1 PP for every 2 Pp for every 1 pp. In other words, a *genotypic ratio* of 1:2:1. You can also express the proportion of genotypes as fractions: ¼ PP, ½ Pp, ¼ pp.

4. **Use what you know about the relationship between the alleles to figure out the phenotype for each offspring in the F_2 generation.**

    For flower color in peas, the purple allele is dominant to the white allele. (You know this because the original cross between purple and white parentals produced all purple offspring in the F_1.) So, F_2 offspring with the genotype PP or Pp will be purple, while F_2 offspring with the genotype pp will be white.

  The ratio of phenotypes predicted in the F2 is 3 purple offspring for every one white offspring — in other words, a *phenotypic ratio* of 3:1. Anytime you see a phenotypic ratio of 3:1, it suggests a monohybrid cross where the relationship between the alleles is simple dominance. You can also express the proportion of phenotypes as fractions: ¾ purple to ¼ white.

Punnett squares represent predictions about the result of a cross. The actual results, recorded from observations of the offspring, may differ from the predictions. For example, the Punnett square in Figure 12-1 predicts a 3:1 phenotypic ratio for plants with purple:white flowers. Mendel's actual data was 705 purple-flowered plants and 224 white-flowered plants, which is a ratio of 705:224 or 3.15:1. The Punnett square predicts outcomes based on a completely even mixing between different gamete types. In real matings, this perfect matching doesn't always occur because each fertilization of one egg and one sperm is a random event. The more offspring produced in a mating, the closer the actual observations will be to the prediction made in a Punnett square.

Tracking the Inheritance of Two Independent Genes

By carefully studying the outcome of a cross involving a single gene, Mendel revealed many of the fundamental principles of genetics. But obviously, peas and other organisms have many, many genes — rice plants, for example, have over 40,000 genes and people have about 22,000. So what does inheritance look like when you consider more than one gene at a time? Obviously, the more genes you try to track at once, the more complicated it gets! But by tracking two genes at once, Mendel discovered another fundamental principle of genetics that applies to the inheritance of multiple genes.

The *Law of Independent Assortment* says that allele pairs for two or more different genes segregate independently during meiosis and the gametes produced combine randomly.

You can best understand the full meaning of this law, and how Mendel discovered it, by studying a *dihybrid cross*, a mating between two individuals who are heterozygous for two different traits.

Adding plant height to the mix

One characteristic of pea plants that Mendel studied was plant height. Pea plants may be tall, or dwarf, which is very short. When pure-breeding tall plants are crossed with pure-breeding dwarf plants, the F_1 offspring are all tall, showing that tall is dominant to dwarf. A monohybrid cross between the F_1 results in a 3:1 ratio of tall:dwarf in the F_2 generation, indicating that a pair of alleles controls this trait and that their relationship to each other is one of simple dominance.

To study the inheritance of two traits at once, Mendel crossed parental plants that were pure-breeding for two characteristics. For example, Figure 12-2 shows a cross between tall plants with purple flowers and dwarf plants with white flowers. The F_1 offspring are all tall plants with purple flowers.

Part III: Making More Plants: Plant Reproduction and Genetics

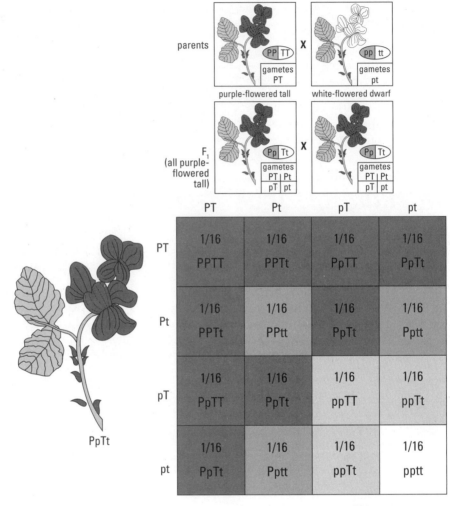

Figure 12-2: A dihybrid cross.

Based on the results in the F1, you can begin to set up the notation for the cross:

1. **Tall is dominant to dwarf and purple is dominant to white.**

   To represent the gene that controls plant height, you can use T for the tall allele and t for the dwarf allele. To represent the gene that controls flower color, you can use P for the purple allele and p for the white allele.

2. **Each plant has two copies of each gene.**

   You can begin to assign genotypes with the P_1 generation. Parents are pure-breeding, so they must be homozygous for both traits. That makes the tall purple parent's genotype PPTT and the dwarf white parent's genotype pptt.

3. **Parents give one copy of each gene to offspring.**

   Each parental must give one copy of the height gene and one copy of the flower color gene to its offspring. Because the parentals are homozygous, they can give only one type of allele for each gene. Tall purple parents give a tall allele (T) and a purple flower allele (P) to their offspring, while dwarf white parents give a dwarf allele (t) and a white flower allele (p). When sperm and eggs combine, they produce the F1 generation, which are all dihybrids (PpTt).

The next step is to figure out what happens when you cross two of the dihybrids from the F1 generation.

Solving the Punnett square for a dihybrid cross

The dihybrid cross follows the same rules as the monohybrid cross: Parents have two copies of each gene, and they must give one copy of each gene to their offspring (by giving one copy of each gene to gametes). What makes the dihybrid cross more challenging is that you have think about two different genes when you're assigning genotypes to gametes.

Dihybrids are heterozygous for both genes, so they have two possibilities for each gene when you consider which alleles they'll give to gametes. For example, the F1 generation shown in Figure 12-2 are all tall plants with purple flowers that have the genotype PpTt.

Dihybrid plants can make four different types of gametes:

- A gamete with a purple allele and a tall allele (PT)
- A gamete with a purple allele and a dwarf allele (Pt)
- A gamete with a white allele and a tall allele (pT)
- A gamete with a white allele and a dwarf allele (pt)

Anytime you're doing a cross involving two different genes, you can use FOIL to figure out the gamete types. FOIL stands for first, outer, inner, last. When you look at the two pairs of alleles for two genes (like in a dihybrid that's AaBb), take the first member of each pair (the A and the B), the outside member of each pair (the A and the b), the inside member of each pair (the a and the B), and the last member of each pair (the a and the b). If you remember FOIL, you'll always make sure to get all possible gamete types. In this example, they're AB, Ab, aB, and ab.

Once you've figured out the gametes produced by the dihybrid parents, you set up the Punnett square just as you would for a monohybrid cross:

1. **Write the gamete types produced by each parent along the edges of the Punnett square.**

   Because dihybrids make four types of gametes, the Punnett square will be a 4x4 square, with a total of 16 boxes.

2. **Bring the possible gametes together, writing the combination in the box at the intersection of the two gamete types.**

   The combinations of alleles represent the predicted genotypes for the F_2 generation.

3. **To figure out the genotypic ratio in the F_2., make a list of the different genotypes and count up how many examples you find of each genotype.**

   For the cross shown in Figure 12-2, the possible genotypes are PPTT, PPTt, PPtt, PpTT, PpTt, Pptt, ppTT, ppTt, and pptt. When you count up the examples, you'll find 1 PPTT, 2 PPTt, 1 PPtt,, 2 PpTT, 4 PpTt, 2 Pptt, 1 ppTT, 2 ppTt, and 1 pptt.

   The genotypic ratio for a dihybrid cross of two genes with simple dominance is 1:2:1:2:4:2:1:2:1.

4. **Figure out the phenotypic ratio in the F_2.**

   Using what you know the relationship between the alleles — in other words, which one's dominant and which one's recessive — figure out the phenotypes that results from each genotype. For the cross in Figure 12-2, the phenotypes from each genotype are purple tall from PPTT, PPTt, PpTT, and PpTt; purple dwarf from PPtt, and Pptt; white tall from ppTT and ppTt; and white dwarf from pptt.

Then make a list of the different phenotypes and count up how many examples you find of each one. In Figure 12-2, the F2 generation has four different phenotypes: tall plants with purple flowers, tall plants with white flowers, dwarf plants with purple flowers, and dwarf plants with white flowers. When you count up how many plants are predicted for each type, you get 9 tall purple, 3 tall white, 3 dwarf purple, and 1 dwarf white.

The phenotypic ratio for a dihybrid cross of two genes with simple dominance is 9:3:3:1.

The Punnett square for a dihybrid cross predicts the types and ratios of offspring that you would see in the F2 generation. Actual observed traits may differ from these predictions. For example, what if you crossed two dihybrid plants and produced only 12 offspring? That's not enough individuals to get ratios close to the predictions, which is why geneticists — and Mendel — need to do many crosses and produce lots of offspring. The more offspring you produce, the more likely it is that the observed ratios will be close to the predicted ratios.

When Mendel did dihybrid crosses, he didn't know about genes and chromosomes. He did the crosses, figured out the ratios, and then used those ratios to build an understanding of what was happening in a dihybrid cross. He figured out that the 9:3:3:1 ratio would result only if the "factors" controlling the two different traits, such as plant height and flower color, worked independently from each other so that dihybrids produced four types of gametes. Since Mendel's time, scientists discovered the process of meiosis and figured out just how cells make gametes. And by studying meiosis, scientists figured out discovered the cellular events behind Mendel's Law of Independent Assortment.

Remembering meiosis

Independent assortment of genes occurs during meiosis I (see Chapter 11). The genes that control two independent genes are located far away from each other — for example, on two completely different chromosomes. For example, the gene for flower color may be on one chromosome, while the gene for plant height may be on a different chromosome. When homologous pairs of chromosomes line up during metaphase I, each pair of homologous chromosomes lines up independently from the others.

So, when a dihybrid plant makes the spores that lead eventually to gametes, meiosis can happen two different ways with regards to the two genes being studied, as shown in Figure 12-3:

- **Meiosis in one cell may lead to both dominant alleles lining up on the same side of the cell equator.** Both dominant alleles would travel to the same side of the cell, making gametes with both dominant alleles (for example, "PT"). Likewise, both recessive alleles would travel to the other side of the cell, making gametes with both recessive alleles (for example, "pt").

- **Meiosis in another cell may lead to the dominant allele for one gene on the same side of the cell equator as the recessive allele for the other gene.** One dominant and one recessive allele would travel to each side of the cell, making gametes that have one dominant and one recessive allele (for example, "Pt" and "pT").

Independent assortment of homologous chromosomes during meiosis I is the reason that dihybrids make four types of spores (that lead to gametes).

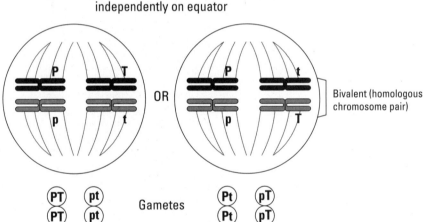

Figure 12-3: Independent assortment of homologous chromosomes during meiosis leads to different combinations of alleles in gametes.

Mixing it Up with Incomplete Dominance

Some alleles don't completely hide the effect of other alleles in an organism that's heterozygous for the two alleles. Scientists call alleles that only partially mask the effect of other alleles *incompletely dominant or codominant*. Incomplete dominance results in heterozygotes that have a phenotype halfway between the phenotypes of the two homozygous phenotypes, and codominance results in heterozygotes that have a phenotype that mixes characteristics of the two homozygotes. (They're essentially the same thing on a molecular level.)

Chapter 12: Passing Plant Characteristics to the Next Generation 213

In snapdragons, for example, inheritance of flower color can have a different pattern than you've seen in peas:

- **The allele for red color is incompletely dominant over the allele for white color.** Because the relationship between these two alleles is not simple dominance, scientists sometimes use a different type of notation, like the one shown in Figure 12-4. One letter is used to represent both alleles, but the variation between the alleles is represented by a super-scripted letter:

    - The allele for red color can be represented as C^R, where C stands for color and the "R" indicates red.
    - The allele for white color can be represented as C^W, where W stands indicates the white allele.

- **Heterozygotes show a phenotype that looks like a blend of the two alleles.** Snapdragons that are homozygous for the red allele ($C^R C^R$) have red flowers. Snapdragons that are homozygous for the white allele ($C^W C^W$) have white flowers. Snapdragons that are heterozygous ($C^R C^W$) have pink flowers.

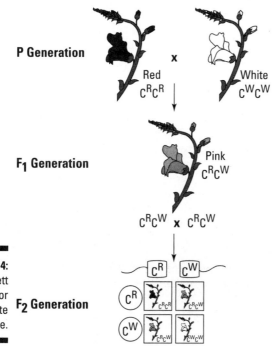

Figure 12-4: A Punnett square for incomplete dominance.

Incomplete dominance leads to different phenotypic ratios in genetic crosses than those caused by simple dominance. As an example, Figure 12-4 presents a cross between snapdragons that have red and white flowers:

1. **When red parentals are crossed with white parentals, all the plants in the monohybrid F_1 generation appear pink.**

2. **A monohybrid cross between individuals within the F_1 generation leads to three possible phenotypes and three possible genotypes in the F_2 generation.**

    The predicted phenotypic ratio is 1 plant with red flowers : two plants with pink flowers : 1 plant with white flowers, or 1:2:1.

    The predicted genotypic ratio is 1 homozygote for red flowers ($C^R C^R$) : 2 heterozygotes ($C^R C^W$): 1 homozygote for white flowers ($C^W C^W$), or 1:2:1.

A monohybrid cross with alleles that are incompletely dominant yields a phenotypic ratio of 1:2:1, which is the same as the genotypic ratio.

Part IV
The Wide, Wonderful World of Plants: Plant Biodiversity

In this part . . .

Plants appeared on Earth long ago and have evolved many complex and wondrous forms, from tiny floating pond weeds to the mighty redwood tree. Just as you can draw a family tree for your family, scientists draw family trees, that show relationships among all the living things on earth. In this part, I introduce the amazing diversity of plant groups found in the family tree of life on Earth.

Chapter 13

Changing with the Times: Evolution and Adaptation

In This Chapter
▶ Understanding the sources of genetic change
▶ Examining plant evolution
▶ Checking out plant adaptations

*L*ife on Earth is constantly changing in response to environmental changes. Life, including plants, migrated from the oceans to the land, dinosaurs have come and gone, and modern humans evolved and spread over the face of the Earth. Biological evolution is the process that leads to changes in the species of life on Earth. Biological evolution occurs through a combination of genetic changes and natural selection. This chapter presents the fundamentals of biological evolution, examples of how to measure evolution in populations, and an exploration of some of the amazing adaptations resulting from plant evolution.

Figuring Out the Fundamentals of Evolution

Evolution is change that occurs over time. *Biological evolution* more specifically refers to changes in living organisms that occur over time. Life on Earth is constantly changing, usually in ways so small you'd hardly even notice. But if you look over huge spans of geological time — millions or billions of years — you can see the big changes that result from biological evolution, from the migration of life from the ocean to the land to the rise and fall of the dinosaurs. During this history of life on Earth, plants have changed from strictly aquatic organisms to simple land plants that reproduced by spores to the more dramatic cone-bearing and flowering plants that grow all around you today.

Mutation

Evolution is possible because living things have a built-in engine for change. Whenever cells copy their DNA so they can reproduce, the enzyme that copies their DNA makes a few small mistakes called mutations. Scientists call the mutations that normally occur during the copying of DNA *spontaneous mutations*.

Most mutations either have no effect or have a negative impact on the organism. Some mutations lead to significant problems, such as genetic diseases. But rarely, a mutation causes a difference that is not negative and may even give an organism an advantage. When you consider the long time span of evolutionary time, even rare events become possible.

Some environmental factors, such as certain chemicals or types of radiation, can increase the rate at which mutations occur in a cell. Scientists call mutations that are caused by exposure to something outside the cell *induced mutations*. The environmental agents that cause mutations are *mutagens*.

Scientists may use mutagens to deliberately induce mutations in organisms that they're studying. For example, scientists create mutations in the mustard weed *Arabidopsis thaliana,* study the mutant plants in order to figure out which genes are mutated, and then discover the normal function of those genes. *A. thaliana* is a flowering plant that has many genes in common with other flowering plants, so by figuring out how *A. thaliana* functions, scientists also learn something about plants that have economic importance to humans. These discoveries can lead to improvements in yield and pest resistance in crop plants.

Natural selection

Mutations cause changes in individuals or their offspring, but evolution is changes in populations over time. In order for changes in populations to occur, other factors must influence the relative success of individuals within the population.

Natural selection is the theory that, in any given generation, some individuals are more likely to survive and reproduce than others due to their genetic traits. People often refer to this theory as "survival of the fittest." Basically, natural selection means that individuals that have the most advantageous traits in any given circumstance are more likely to survive and reproduce, passing their traits onto their offspring. So, genetic traits that give an advantage will tend to increase in a population, causing that population to change over time.

Chapter 13: Changing with the Times: Evolution and Adaptation

The characteristics of organisms that make them suited to their environment are *adaptations*.

Observing the natural world with Charles Darwin

Natural selection was first proposed by Charles Darwin, an English naturalist who made detailed observations both at home and on his travels around the world. Darwin made several observations that led him to propose his idea of *descent with modification* — in other words, living things descend from their ancestors, but they can change over time:

- ✓ **Artificial selection occurs when people use selective breeding to change populations of farm animals, dogs, crop plants, and other species to change over time.** When farmers breed the sweetest corn plants together or hunters choose their best hunting dogs to produce puppies, they're increasing the chances that the traits they desire will be present at higher numbers in the offspring.

- ✓ **Extinction occurs when all members of a particular species die out.** Along a riverbed in Argentina, Darwin discovered the fossil bones of three extinct species. The extinct species resembled three living species, so Darwin began to wonder whether the prior species were ancestors to current species and whether all existing species were somehow connected to species in the past.

- ✓ **Adaptive radiation happens when members of one species become specialists for different types of environments, causing groups of specialists to become new species over time.** Darwin observed adaptive radiation when he traveled to the Galapagos Islands, an island chain off the coast of South America. Darwin particularly focused on a type of bird, called finches, that lived on the Galapagos Islands. Each island had its own unique species of finch that were distinct from each other and from the finches on the mainland. In South America, this type of finch ate only seeds. On the islands, some finches ate seeds, others ate insects, some even ate cactus. The beak of each type of finch seemed exactly suited to its food source. Darwin thought that all the finches had a common ancestor from mainland South America that either flew or floated to the newly formed islands, perhaps during occasional storms. Once the ancestral finches reached the islands and began reproducing, different groups became specialists on unique types of foods. Darwin thought that within each group, the finches that had the beaks that were best able to use that particular food would be more likely to reproduce and pass their type of beak onto offspring. Over long periods of time, the beaks of different groups of finches became distinct from each other.

When a population becomes so distinct in its characteristics that it can no longer successfully mate with related populations, it is a new *species*. When new species are created, the process is called *speciation*.

Putting the theory together

Before he finally published them, Darwin thought about his ideas on evolution for more than 20 years, partially because he was worried about the reaction of the church to his ideas. During that time, Darwin worked on his ideas, gathering evidence from agriculture, geology, and the natural world to support his thinking. Finally, when Darwin heard that another scientist, named Alfred R. Wallace, was about to publish similar ideas, Darwin felt compelled to share his work with the world.

In his book *On the Origin of Species by Means of Natural Selection,* Darwin presents the fundamental requirements for natural selection to occur:

- **Individuals in most natural populations produce more offspring than can survive.** A single maple tree produces thousands of maple seeds each year, very few of which actually survive to grow into new maple trees. Each seed must compete with the others for space, light, water, and nutrients.

- **Individuals have unique characteristics.** During Darwin's time, no one knew where these differences came from. Now scientists know that differences in organisms arise due to mutations in DNA and the mixing of genetic information that occurs during sexual reproduction.

- **Some characteristics are inherited from parents to offspring.** During Darwin's time, the laws of inheritance were just beginning to be figured out, so Darwin didn't know exactly how parents passed on their traits. Now scientists know that traits are inherited when parents pass genes onto their offspring. (For more on inheritance, see Chapter 12.)

- **Organisms with characteristics best suited to their environment are more likely to survive and reproduce.** This idea is the heart of natural selection: If organisms must compete to survive and all organisms aren't the same, then the ones with the advantageous traits are more likely to survive — the fastest growing maple seed perhaps. If these traits can be inherited, then the next generation will show more of these advantageous traits. An example of natural selection is shown in Figure 13-1, where a hawk is shown looking for mice to eat. Hawks are visual predators, so they are more likely to attack the mice that are easiest to see. In the figure, light-colored mice shown up better against the background of grass and dirt, so the hawk is more likely to eat them. Over time, the population of mice that lives in this environment is likely to have more darker colored individuals that result from the reproduction of dark-colored survivors. Predation like this can also influence the evolution of plant populations. A browsing herbivore may enjoy eating the individuals of a particular plant species. If some of the individuals in the population had rougher, almost spiny edges on their leaves, the herbivore might avoid these individuals and graze on the softer members of the population. The spiny-edged individuals would survive in greater numbers, producing offspring that also had spiny edges. Over time, the entire population of this plant species might become spiny-edged.

The environmental factor that causes some organisms to survive and others to not survive is called a *selection pressure*. (Because it puts "pressure" or stress on the individuals of the population.)

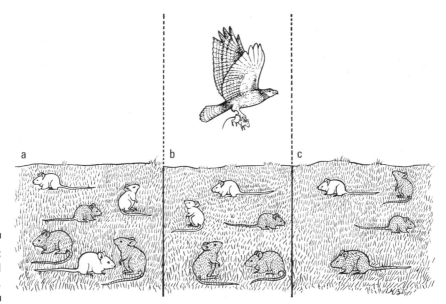

Figure 13-1: Natural selection.

It's very important to realize that biological evolution happens to populations, not to individuals. Individuals live or die, reproduce or don't, depending on their circumstances. But individuals themselves can't evolve in response to a selection pressure. Imagine a plant growing in the forest under some trees so that it doesn't get quite enough light. The individual plant can't suddenly change itself into a vine that can climb up a tree trunk to get to the light in the forest canopy. But if an individual has a mutation so that some of its leaves are like tendrils that can wrap around things, maybe that individual will be able to maneuver itself into a lighter spot. If some of its offspring have leaves that are even more tendril-like, they may become more successful. Over long periods of geologic time, where natural selection favors the individuals with tendril-like leaves, the plant population may evolve into climbing vines.

Identifying Important Factors in Plant Evolution

Many of the processes that drive evolution, such as mutation, natural selection, migration, and genetic drift, can occur in populations of all types of organisms. But for each type of organism, certain processes may have greater impacts than others. Plants, for example, don't migrate as easily as do animals. Animals, on the other hand, aren't as likely to produce fertile *hybrids,* individuals that are the offspring of two different species (or varieties), as are plants.

You're probably familiar with the example of a horse and a donkey mating to produce a mule. Mules function normally, but can't produce offspring of their own, so they represent an evolutionary dead end. Many plant species, on the other hand, can mate with other species and produce hybrids that are capable of reproducing, potentially leading to new species. Several processes, including hybridization, have made significant contributions to plant evolution.

Hybridization

Most plant hybridization has probably been manmade as a result of human agriculture. As farmers seek to develop crop plants with the most desirable features, they often mate two different varieties of a plant, or even two entirely different species, in order to try and capture certain qualities from each.

When different varieties or species of plants are mated to each other, the F_1 generation is often bigger and healthier than either parent plant. Scientists call the improved health of hybrids *hybrid vigor*.

Taming the wild grasses

A major cultural shift in the human population occurred about 10,000 years ago when people switched from hunting and gathering to farming. Wheat, which is a grass, was one of the first plants to be domesticated. Studies of chromosome number in wild and cultivated varieties of wheat have pieced together the probable evolution of modern wheat varieties. One of the first wild wheat species to be cultivated widely was *Triticum monococcum*, a wheat species with 14 chromosomes (7 homologous chromosome pairs). Emmer wheat, *T. turgidum,* was another very early cultivar and was also gathered by humans of the Paleolithic period (or Stone Age). *T. turgidum* has 28 chromosomes (14 pairs) and probably evolved through gene doubling from *T. monococcum* or a related hybrid.

Most of the species of wheat and corn that farmers grow today are hybrid varieties.

Polyploidy

About half of all flowering plant species are polyploid. Two different processes seem to have caused polyploid species during plant evolution:

- **Gene doubling by failure of chromosomes to separate during meiosis.** During meiosis, the chromosome number of a cell is supposed to be cut in half in order to produce gametes. *Nondisjunction* is the failure of chromosomes to separate properly during cell division. If nondisjunction occurs, then gametes may be produced that have twice the normal number of chromosomes. If these gametes go on to fertilize other gametes, they can produce polyploid plants.

- **Hybridization followed by gene doubling.** Hybridization may produce hybrids with mismatched sets of chromosomes. These hybrids can't reproduce because they can't successfully complete meiosis. But if a cell of one of these infertile hybrids tries to divide but then fails to separate the chromosomes, it could produce a cell with two complete sets of the originally mismatched chromosomes — in other words, a cell with two matching sets of chromosomes! This cell can now undergo a normal meiosis, producing gametes that can lead to new offspring. (Hybridization and gene doubling appear to have led to the evolution of modern wheat; see the sidebar "Taming the wild grasses" for details.)

Reproductive isolation

Reproductive isolation, which is the separation of two populations so that they can't interbreed, can occur in many different types of organisms. When two populations become reproductively isolated, they can't mix their alleles anymore, and they may begin to evolve independently from each other. Eventually, they may become so different that they can no longer interbreed and have become distinct species from each other. Scientists call the process *divergent evolution.*

Admiring Plant Adaptations

Some of the truly amazing results of plant evolution are the ways in which plants have become specialized to survive and take advantage of the conditions in different habitats. When you consider all the amazing strategies plants have evolved for survival, it seems like there's almost no challenge they can't overcome. As examples of the incredible power of plant evolution to create diversity, some very cool plants have evolved to meet the challenges of unique environments.

Desert plants

Life in the desert can be difficult for a plant — deserts are dry, have extremes of temperature (hot during the day, but cold at night), and can have almost too much sunlight. These conditions put strong selection pressures on plants that have strategies to live with low water and protect themselves from dry air and bright light. Many different species of plants live in the desert, but even very unrelated species have evolved similar strategies for dealing with desert conditions. Sometimes, two plants that look very much alike because they've adopted the same strategies are actually completely unrelated to each other.

When two genetically unrelated species evolve to look similar because of similar selection pressures, scientists call the process *convergent evolution*.

Figure 13-2 shows some common strategies plants use to survive desert conditions:

- **Reduced or highly modified leaves:** Many plant leaves are thin and flat, ideal for catching sunlight, but not at all suitable for desert life. Those thin flat surfaces would be exposed to lots of dry air, causing water to evaporate rapidly out of the stomates.

  - In many desert plants, such as the barrel cactus in Figure 13-2, leaves have evolved into spines, which don't lose water rapidly and which protect the plant from grazers. These plants have switched the main site of photosynthesis to the stem.

  - Other desert plants, such as the agave shown in Figure 13-2, have evolved tough, thick leaves that have a heavy wax coating to prevent water loss.

  - Some plants, such as the desert yam shown in Figure 13-2, produce delicate, thin leaves, but they only make them during the rainy season when it's safe! The rest of the year, the leaves die back only the plants' tough, protective base remains.

- **Organs for water gathering and storage:** Many desert plants are *succulents,* plants that store water in fleshy tissue. All four desert plants shown in Figure 13-2 have ways of dealing with the desert water shortage. Cacti store water in their stems, which have pleats to allow expansion when water is available. Agave plants have long root systems that can find water under the desert ground. The desert yam stores water in its tough base, and stone plants have succulent stems.

Chapter 13: Changing with the Times: Evolution and Adaptation 225

a. Spiny leaves and succulent, photosynthetic stems (barrel cactus)

b. Succulent leaves (agave)

c. Seasonal leaf production (desert yam)

d. Reduced exposure to light (stone plants)

Figure 13-2: Common strategies plants use to survive desert conditions.

- **Protection from overexposure:** Too much light or exposure to dry air can damage a plant, so desert plants have evolved structures and strategies to protect themselves from these conditions. Some cacti have very long, white, almost hair-like spines form a surface barrier over the stem, slowing down air movement and reducing damage from overexposure. Because the spines are white, they also help reflect excess sunlight off the stem. Some plants, such as the stone plant shown in Figure 13-2, snuggle themselves down into the desert floor, allowing just their surfaces to come into contact with the elements. The surface of stone plants have little clear spots that allow sunlight to enter the plant, as if it were passing through a window so that photosynthesis can occur inside the submerged stem.

Tropical rainforest plants

The environment of a tropical rainforest is almost the opposite of that of a desert — tropical rainforests are wet, and plants can have a hard time getting enough light. Light is a problem for many plants because the trees of the tropical rainforest spread their big leaves to form a canopy and block light from reaching the plants down below. Another environmental stress in the tropical rainforest is that the soil is not very deep and of poor quality.

These selection pressures caused populations of rainforest plants to evolve unique features that help them succeed in their environment:

- **Prop, stilt, and buttress roots:** Rainforest soil is actually so shallow that rainforest trees can't put down deep enough roots to anchor themselves properly. So, many rainforest trees make adventitious roots that help provide additional support. (For more on plant roots, check out Chapter 4.) The prop roots shown in Figure 13-3 are one example of adventitious roots made by rainforest trees.

- **Waxy leaves with drip tips:** Too much water can actually be a problem in the rainforest — if water sits on a plant leaf, the leaf could get heavy and break, or pathogens such as fungi might begin to grow and attack the leaf. So, many rainforest plants have extra wax and a pointed tip on the leaf that help the excess water slide off.

- **Structures for clasping and climbing:** Because of the big leaves on the rainforest trees, the only place in the rainforest that gets lots of light is the upper canopy. The rainforest is full of different kinds of plants that have evolved ways to grow where the action is:

  - *Vines*, such as the liana shown in Figure 13-3, put roots into the soil, but then wrap around other plants in order to grow upwards toward the light.

  - *Epiphytes* are plants that grow entirely on other plants, without putting roots into the soil at all. Many tropical orchids are epiphytes, as are the bromeliads like the one shown in Figure 13-3. Because epiphytes have no roots in the ground, they have to have strategies for getting water and minerals up in the canopy. Orchids have adventitious roots covered with spongy material that helps them collect water and particles from the air. Bromeliads are also called tank plants because they form a central water storage area by wrapping their leaves tightly into a cylinder — basically forming a little pond in the center of the plant. Up in the rainforest canopy, tank plants not only catch water for themselves, but they also form homes for little animals, such as tree frogs.

Chapter 13: Changing with the Times: Evolution and Adaptation 227

a. Prop roots

b. Waxy leaves and drip tips (philodendron)

Figure 13-3: Adaptations of rainforest plants.

c. Epiphytes (Bromeliad)

d. Vines (liana)

Carnivorous plants

Carnivorous plants grow in different environments that all have one thing in common: low nutrient availability. Many carnivorous plants grow in acidic environments, such as peat bogs, where the high acid content makes it difficult for plants to take up minerals through their roots. Some carnivorous plants grow as epiphytes in the canopy of tropical rainforests, and so don't put roots into the soil at all. Whatever the reason for the shortage of minerals, carnivorous plants evolved the same solution — catch bugs and absorb the minerals from their rotting corpses!

Carnivorous plants do photosynthesis just like all other plants and can make their own sugars from carbon dioxide and water. Thus, they don't need to capture prey to supply themselves with energy or with carbon, hydrogen, or oxygen — they capture prey to supply themselves with minerals, such as nitrogen and phosphorous, that they are unable to get from the soil.

Carnivorous plants have evolved many different strategies for catching their prey, including

- **Modified leaves that snap closed to trap prey:** The Venus flytrap, shown in Figure 13-4, is probably the most famous of all carnivorous plants. Its traps are modified leaves, which are covered with hairs that act as triggers to spring the trap. When an insect crawls across the surface of the trap and touches a hair, not just once, but twice, the trap snaps shut to capture the prey. The plant releases digestive juices that contain enzymes to help speed up decomposition and release of minerals.

- **Modified leaves to form pitchers that trap prey:** Pitcher plants, such as the one in Figure 13-4, have modified leaves that form tall columns for trapping prey. (For the details on how pitcher plants, such as the cobra lily, trap and kill their prey, see Chapter 21.)

- **Glands that produce sticky substances to trap prey:** Plants such as the sundew, shown in Figure 13-4, have elongated glandular structures that secrete a sticky, sweet mucus. The sweet mucus attracts insects that then get trapped in it and either die from exhaustion as they try to escape or drown in the mucus itself. The glands also secrete digestive enzymes that break down the body of the prey, releasing minerals to the plant.

Chapter 13: Changing with the Times: Evolution and Adaptation   **229**

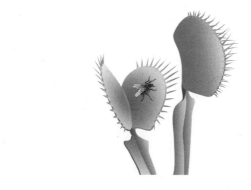

a. Venus fly trap

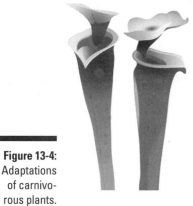

b. Pitcher plant

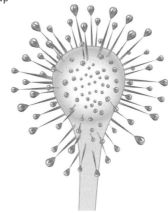

c. Sundew

Figure 13-4: Adaptations of carnivorous plants.

Aquatic plants

Aquatic plants obviously have access to plenty of water and dissolved minerals, so their challenge is to find a way to balance life in the water with access to sunlight. Evolution has led to plants (see Figure 13-5) that utilize a variety of strategies for making life on the water work:

- **Flat, floating leaves:** Water lilies are a familiar example of this strategy. Water lilies are rooted in the sediment, but their leaves have very long petioles that stretch to the surface of the water. The leaves themselves are flat so that they can float easily. The petioles contain aerenchyma tissue with lots of air spaces to help the leaves float. (For details on aerenchyma, check out plant tissues in Chapter 3.)

- **Buoyant tissues:** Water hyacinths have bulbous, spongy petioles that are full of aerenchyma. The petioles are so buoyant the whole plant floats. The short roots of water hyacinth extend into the water, but don't anchor the plants in the sediment.

a. Water lily

Aerenchyma

b. Water hyacinth

c. Duck weed

Figure 13-5: Adaptations of aquatic plants.

> ✓ **Small size, rounded shape:** The aquatic plant duckweed is the smallest and fastest growing of all the flowering plants. They have no true leaves or stems, but are just small rounded bodies that float on the water.

Chapter 14

The Tree of Life: Showing the Relationships Between Living Things

In This Chapter

▶ Discovering the three domains of life
▶ Reading phylogenetic trees
▶ Naming organisms
▶ Using a dichotomous key

*L*ife on earth is wonderfully abundant and diverse. People, such as scientists, farmers, gardeners, and nature lovers, who are interested in living things look for ways to organize what they know. An important part of that organization is naming each organism and observing the characteristics that make each organism unique. This chapter introduces the methods that scientists use when naming and classifying organisms and gives examples of how to read phylogenetic trees and use dichotomous keys.

Examining the Branches of the Tree of Life

Chances are the members of your family have more things in common with each other than they do with people who aren't related to you — you may have similarities in eye and skin color, a certain shape to your noses, or a tendency to be tall. You have more in common with your relatives than you do strangers because your family all share certain genes in common. You got your DNA from your parents, who got theirs from their grandparents. Some of this same DNA got passed to your siblings, cousins, aunts, and uncles, leading to your *shared characteristics,* the characteristics that you have in common.

In the same way that you can recognize similarities between your family members, you can probably pick out relationships between other kinds of living things on earth. Cats and dogs may be different, but they both have fur, four legs, and a tail and are clearly more like each other than they are like a rosebush. So, if you had to draw a family tree of life on earth, you'd probably put cats and dogs fairly close together on a branch of the tree and pretty far away from the rosebush branch. Scientists use this same basic method of looking for similarities between living things in order to figure out the relationships between all life on earth.

Scientists assume that the more shared characteristics two organisms have with each other, the more closely related they are to each other.

The types of clues scientists use to figure out relationships include

- **Physical structures:** The structures that scientists use for comparison may be large, like the production of cones versus flowers, or very small, like the presence of a wall around the cell. For example, cone-bearing plants are more closely related to each other than are flowering plants. Scientists consider reproductive structures like cones and flowers to be especially important for determining relationships.

- **Chemical components:** Some organisms make unique chemicals. All plants, for example, make cell walls made of cellulose and produce starch as their main energy storage molecule.

- **Genetic information:** An organism's DNA determines its traits, so by reading the DNA, scientists go right to the source of differences between species. Even organisms that might seem to be incredibly different — such as you and a carrot — have traits in common.

Digging into the three domains

All cells on earth contain ribosomes for making proteins. So, scientists can read the genes that contain the blueprints for ribosomes from any type of cell on earth and use the similarities and differences between the genes to build a family tree for all life on earth. The genes scientists use most often for looking at relationships are called *ribosomal RNA genes* (rRNA genes).

When scientists compared all life on earth by reading the codes of ribosomal RNA genes, they discovered that living things fall into three main groups. Scientists call the three main groups of life *domains*.

Figure 14-1 shows the three domains of life that make up the main branches of the family tree of life on earth:

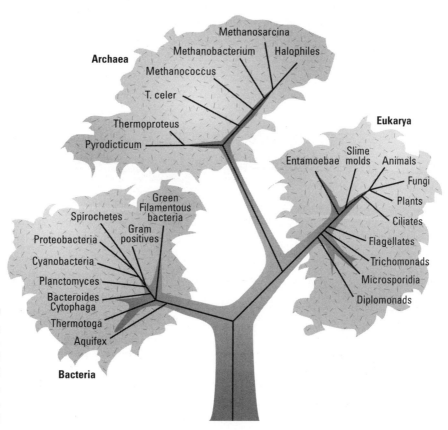

Figure 14-1: A phylogenetic tree of life based on rRNA genes.

- ✔ **Archaea:** Archaeans are prokaryotic organisms that are found in just about every environment on earth, including harsh environments like hot springs, salt lakes, and hydrothermal vents. (For the scoop on prokaryotic cells, check out Chapter 2.) Although archaeans physically resemble bacteria, the two groups of prokaryotes have some fundamental chemical and genetic differences that indicate that they're unique types of cells.

- ✔ **Bacteria:** Bacteria are prokaryotic organisms that mostly get noticed by people because some of them cause disease. Although some bacteria are truly dangerous to human health, most bacteria are actually beneficial to humans and all other types of life on earth. Bacteria are nature's recyclers — they break down dead material and recycle nutrients that other organisms need to live. Many bacteria are plant-like in the sense that they're photosynthetic — in fact, bacteria and algae perform over half the world's photosynthesis!

✔ **Eukarya:** Plants and people — as well as other animals, fungi, and lots of microorganisms — belong in this domain. All organisms in Eukarya have eukaryotic cells. (For details on the eukaryotic cell, flip to Chapter 2.) So, even though you're physically very different from plants, they're your very distant relatives!

Exploring the past through phylogenetic trees

The trees that scientists draw to represent relationships between living things, like the ones in Figures 14-1 and 14-2, are called *phylogenetic trees*. Scientists use computer programs to draw the trees based on their observations of structure, chemistry, and/or gene sequences of individual species or related groups called taxa.

A *taxon* (plural: *taxa*) is a group of related organisms in a biological classification system.

The computer programs compare the similarities between different species, drawing the trees so that the branches of the most similar groups are closest together and sketching in how the different groups might have descended from their ancestors.

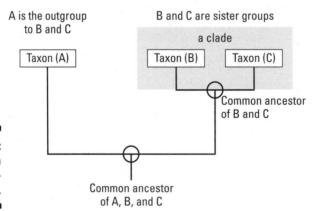

Figure 14-2: Reading a phylogenetic tree.

Phylogenetic trees not only show how closely related organisms are, they also help map out the evolutionary history, or *phylogeny,* of life on earth.

Chapter 14: The Tree of Life: Showing the Relationships Between Living Things

Finding your long-lost relatives

Although scientists have been drawing trees for life on earth for hundreds of years, bacteria presented a bit of a problem. You see, for most of that time, scientists were basing their ideas about relationships on the physical appearance of organisms and bacteria — well, all bacteria look pretty similar. But then, in the 1970s, scientists began developing the technology to read DNA and RNA molecules. Instead of looking at an organism's traits, the door opened to the possibility of reading the DNA blueprints for those traits.

One pioneer in this field was a scientist named Carl Woese who began using the new technology to read the RNA molecules of prokaryotes, which at the time everyone thought were all bacteria. But Woese found something remarkable — when he read the RNA code of prokaryotes, he discovered two distinct groups. One group with similar sequences was in fact, the bacteria. But the other group, which on the surface looked like bacteria, had very different RNA sequences. And when Woese compared the RNA sequences of the bacteria and this new group with sequences from eukaryotes, he found that the new group was just as different from the bacteria as bacteria were from eukaryotes. So, even though the two groups of prokaryotes looked alike, they were actually very distantly related.

Based on his results, Woese proposed that life on earth consisted of three main types, which he called *domains*. Many scientists initially resisted this idea — at the time, everyone thought that there were five kinds of life on earth — bacteria, plants, animals, fungi, and protists. But as more scientists investigated the new group — called Archaea — the evidence supported Woese's idea. Archaeans weren't just genetically different from bacteria, they also had some fundamental chemical differences.

The ripple effects of Woese's work were huge — Woese not only discovered a type of life on earth to which people had been oblivious, he also showed the power of the new technology for reading DNA and RNA and added this powerful new tool to the study of relationships of life on earth. Since Woese's work, scientists in all fields have gone back to re-examine the relationships between organisms, using genetic comparisons to test old ideas and clear up past mysteries — like those of the relationships between groups of plants that look similar to the earliest plant fossils!

Just like your family began a long time ago with your original human ancestors, scientists believe that all life on earth began from one original *universal ancestor* after the earth formed 4.5 billion years ago. Most phylogenetic trees reflect this idea by being *rooted,* meaning they're drawn with a branch that represents the common ancestor of all the groups on the tree. In Figure 14-1, earlier in this chapter, the unlabeled branch at the bottom of the tree represents the common ancestor for all organisms on the tree, which in this case is the universal ancestor of all life on earth.

To read a phylogenetic tree like the one in Figure 14-2, look for the following information:

- The tips of the branches represent the species or other taxa that the scientist is comparing. Figure 14-2 shows a tree for three taxa: taxon A, taxon B, and taxon C.

- Branches meet at points called *nodes* that represent the common ancestor of the two taxa. Scientists call groups that branch out from the same common ancestor *sister groups*. In Figure 14-2, group B and group C are sister groups.

- An ancestor plus all its descendants form a *clade*. Figure 14-2 shows the clade for taxon B and C.

- Scientists call groups that branch from the base of the tree and are separate from the other groups *outgroups*. Scientists often deliberately include observations about a group that isn't very closely related to the group being studied in order to give a tree an outgroup. When the computer program has to include the outgroup, it helps give the tree scale by showing the group being studied in relationship to the larger picture of other kinds of life on earth.

Organizing Life

The three domains represent the three largest, and most distantly related, groups of living things on earth — in other words, the three biggest branches on the family tree of life. But each of those big branches has many smaller branches that represent related groups within the domains. Each living thing is classified by its characteristics into a *taxon*.

Developing a system

Scientists who name and organize living things practice the branch of science called *systematic biology*, or *systematics*. The science of systematics includes subdisciplines called *taxonomy*, which is the naming of living things, and *classification*, which is the organization of living things. Systematic biologists organize life on earth not just by the domains, but by subcategories within the domains. All the categories and subcategories form a *taxonomic hierarchy* that shows the degree of relationship between groups.

The taxonomic hierarchy for living things, from the largest, most inclusive group to the smallest, least inclusive group is Domain, Kingdom, Phylum, Class, Order, Family, Genus, Species.

You can think of the taxonomic hierarchy as a set of nesting boxes. You'd start with three big boxes representing the three domains and sort all the organisms on earth into those three boxes. Then, you'd take one of the boxes — box Eukarya, for example — and look at how you could further organize all the living things you'd put into that box. You'd see that all animals had certain things in common that were different from the rest of the organisms, so you'd put all the animals together in a smaller box inside the big Eukarya box. Similarly, you could put all the plants together in a second smaller box, all the fungi into a third box, and all the protists into a fourth. Then, you could pick up the animal box, look inside and decide how to sort the animals further into even smaller boxes that would fit inside the animal box, and so on, until you'd organized all the living things in each box as much as you possibly could.

In traditional, or *classical systematics,* organisms are sorted into groups primarily based on their similarities in structure (morphology), development, and metabolism. The more similar two organisms are, the more likely they are to be placed together in a category. The relative position of two organisms within the taxonomic hierarchy suggests the degree of their relationship. For example, you and a carrot are both in domain Eukarya, so you definitely have some things in common, but you have more things in common with organisms that are also in the animal kingdom, like a dog. Table 14-1 gives you an idea of how the taxonomic hierarchy works by comparing the taxonomy of you, a dog, a carrot, and the bacterium, *E. coli*.

Table 14-1 A Comparison of the Taxonomy of Several Species

Taxonomic Group	*Human*	*Dog*	*Carrot*	*E. coli*
Domain	Eukarya	Eukarya	Eukarya	Bacteria
Kingdom	Animalia	Animalia	Plantae	Eubacteria
Phylum	Chordata	Chordata	Anthophyta	Proteobacteria
Class	Mammalia	Mammalia	Rosopsida or Eudicots	Gamma-proteobacteria
Order	Primates	Carnivora	Apiales	Enteroacterialess
Family	Hominidae	Canidae	Apiaceae (Umbelliferae)	Enterobacteriaceae
Genus	*Homo*	*Canis*	*Daucus*	*Escherichia*
Species	*Homo sapiens*	*Canis familiaris*	*Daucus carota*	*Escherichia coli*

Of the organisms listed in Table 14-1, you have the most in common with a dog. You're both animals that have a central nervous chord (phylum Chordata), and you're both mammals (class Mammalia), which means you have hair and the females of your species make milk. However, you also have many differences from a dog, including the tooth structure that separates you into the order Primates and a dog into the order Carnivora. If you compare yourself to a plant, you can see that you have certain features of cell structure that place you together in Domain Eukarya, but little else in common. Two organisms that are together in the same species are the most similar of all.

People have lots of funny sayings to help them remember the taxonomic hierarchy. My personal favorite, because it's so hard to forget, is <u>D</u>umb <u>K</u>ids <u>P</u>laying <u>C</u>hase <u>O</u>n <u>F</u>reeways <u>G</u>et <u>S</u>quished. The first letter of each word in the sentence represents the first letter of a category in the taxonomic hierarchy and will help you remember the order of the categories. If you don't like this particular saying, just search the Internet for "taxonomic hierarchy mnemonic," and you'll find many others.

In classical botany, the term *division* was used instead of phylum. Most biologists have switched to using phylum for plants in order to be consistent among all living things, but some botanists still use the older term division.

Now that scientists have the ability to read the DNA of organisms, they are gaining new understandings of the relationships between taxa of life on earth. By going straight to the genes, scientists can compare organisms that don't have a lot of morphological characteristics, like prokaryotes, or see through confusing situations, such as two unrelated organisms having a similar structure. New information is being obtained at a rapid pace, and modern systematic biologists are still working on how best to combine the new information on phylogeny with the old system of using the categories of the taxonomic hierarchy.

Ideally, scientists want to create a taxonomic system that's easy to use and that reflects the evolutionary history of organisms. In order for taxonomic groups to reflect relationships, the groups have to be carefully constructed:

- ✔ **Ideal taxonomic groups are *monophyletic*, meaning that all the organisms within the group share a common ancestor and all descendants of that ancestor are included in the group.** For example, you could create a monophyletic group by grouping taxon A, B, and C all into the same family.

- ✔ **Some taxonomic groups are *paraphyletic*, meaning that all the organisms in the group share a common ancestor, but some of the descendants of that ancestor aren't included in the group.** For example, if you grouped taxon B and C from Figure 14-2 into one family and put taxon A into a different family, then the family containing B and C would be paraphyletic. Both B and C share a common ancestor, but they also share a common ancestor with A, which is being left out of the family. Paraphyletic groups do exist within current taxonomic systems because they sometimes make it easier to discuss certain groups.

- ✔ **Taxonomic groups that are *polyphyletic* (meaning that the organisms within the group don't share a common ancestor but are grouped together because of a common characteristic) should be avoided.** Polyphyly would occur if you grouped the cyanobacteria from domain Bacteria and plants from domain Eukarya into one taxonomic group based on the fact that they're all photosynthetic. Look at Figure 14-1 and imagine drawing a circle around these organisms to create a group, and you'll see that the group makes no sense in evolutionary terms — although cyanobacteria and plants share a common ancestor way back in evolutionary time, putting just them into a group completely ignores many of their closer relatives.

Defining plants

In the classic taxonomic hierarchy, organisms within each domain are broken up into categories called *kingdoms*. Domain Eukarya currently contains four kingdoms: Animalia (animals), Plantae (plants), Fungi (mushrooms, molds, yeast), and Protists (mostly microbial eukaryotes).

Most modern botanists include organisms in Kingdom Plantae if they have the following characteristics:

- **Eukaryotic** (see Chapter 2)
- **Photosynthetic**, with the pigments chlorophyll *a* and chlorophyll *b* (see Chapter 7)
- Store food as **starch** and have **cellulose** in their cell walls (see Chapter 2)
- During cell division, the formation of new cell wall is aided by a **phragmoplast** (see Chapter 11)
- Develop from **embryos** supported by maternal tissue

Based on this definition of plants, combined with historical precedent, the plant kingdom is usually defined as containing land plants, from mosses that don't have vascular tissue (see Chapter 15), to cone-bearing plants (see Chapter 16) and flowering plants (see Chapter 17) that do have vascular tissue.

This definition of the plant kingdom, however, may change as a result of comparisons of the DNA of land plants and the green algae.

Comparisons of genes in land plants and green algae support the idea that green algae are the closest relatives of land plants, something that botanists have believed for centuries based on comparisons of morphology, metabolism, and development.

However, the information from DNA studies also showed that one group of green algae, the charophytes, are a sister group to the land plants. In other words, if you refer to Figure 14-2, you could imagine taxon B as the charophytes and taxon C as the land plants, which are sometimes called *embryophytes*. The charophytes and the land plants share a common ancestor. So, if scientists put the land plants in one kingdom (Plantae) and put the charophytes in another kingdom

(Protista), then the kingdoms are paraphyletic and don't really reflect the evolutionary relationships between these organisms.

Because land plants and green algae form a monophyletic group, many botanists are in favor of including green algae in the plant kingdom.

It's not yet clear how botanists will reconcile the classical definition of the plant kingdom with the more recent understanding of the phylogeny of plants. Some botanists have proposed leaving things as they are, placing charophytes and embryophytes in different taxonomic groups. Others have proposed placing them together in a single phylum called the Streptophyta. In this book, I use the classical definition of plants as stated in this section, but acknowledge the close relationship of charophytes to the land plants.

Naming the Rose (And Other Living Things)

People have been studying, naming, and classifying plants for thousands of years. One early major attempt was made by Theophrastus, a student of Aristotle and Plato. In the third century BC, Theophrastus published a classification of nearly 500 plants that scientists continue to use and add onto until the 18th century, using the structure of leaves, stems, fruit, and flowers to organize and name plants. Related plants were grouped into taxa called *genera* (singular: genus). Each plant was given a name that consisted of the Latin name for the genus, plus a short description, also in Latin, of its distinguishing characteristics. This practice produced some pretty long names, such as *Mentha floribus spicatis, foliis oblongis serrates,* which represented a mint with a spike of flowers, and oblong, saw-toothed leaves.

Fortunately, in the 1750s, a Swedish naturalist named Carl Linnaeus, who did a huge amount of work on classifying plants and animals, streamlined the naming of organisms by proposing a system of naming that included an abbreviated name for each species consisting of just two words. In Linnaeus's system, the abbreviated name (called a *binomial* because it contains only two words) for the mint described in the previous paragraph would be *Mentha spicata*. In time, Linnaeus and scientists after him dropped the longer plant names in favor of the shorter ones, formally adopting a system of binomial nomenclature.

In *binomial nomenclature,* the first part of an organism's name is the genus, and the second part of the species name is the *specific epithet,* or species name.

The rules for using binomial nomenclature are

- The genus is always capitalized.
- The specific epithet is never written without the genus, although the genus can be abbreviated by just the first letter.

Chapter 14: The Tree of Life: Showing the Relationships Between Living Things

✔ To indicate that the name is the official scientific name, both the genus and species are either italicized or underlined.

✔ The name of the person who named the organism may also be included after the name, either as a full last name or the first letter of the last name.

According to these rules, then, humans may be correctly identified as *Homo sapiens* or *H. sapiens*.

All scientists around the world use the same scientific names for organisms, which helps them communicate no matter which language they speak or where they live. Although many plants that have blue bell-shaped flowers might commonly be called bluebells, they may not actually be related — in other words, a bluebell in England may be different from a bluebell in Texas. In order to share information and new discoveries, scientists need to be sure that they're talking about the same organism. So, if you find a bluebell in Texas and call it *Eustoma grandiflorum,* a scientist will know exactly which flower you're talking about.

If you were out walking in a field and you found a plant with a blue bell-shaped flower, you could figure out exactly what to call it using a dichotomous key. *Dichotomous keys* use a series of questions, each of which has two choices, to help a person identify the name of an organism.

Many people who enjoy hiking or bird-watching use dichotomous key to help them identify plants and animals. Some keys designed for recreational use may use other types of organization — for example, grouping all plants with red flowers into one section of a book, and all plants with blue flowers into another. But many keys, especially those used by botanists, rely upon the dichotomous organization shown in Table 14-2.

Table 14-2	A Dichotomous Key	
Numbered Pairs of Traits to Consider	**Description of Plants to Choose From**	**Directions for Next Move**
1a.	Flowers blue to pink, in open or dense clusters	Go to 2
1b.	Flowers white, yellow or blue, in coiled clusters	Go to 3
2a.	Stems ascending to upright; flowers 9–19 mm long	*Mertensia paniculata*
2b.	Stems sprawling; flowers 4–12 mm long	*Mertensia maritime*
3a.	Flowers blue; nutlets round, smooth, and shiny	*Myosotis laxa*

(continued)

Table 14-2 *(continued)*

Numbered Pairs of Traits to Consider	Description of Plants to Choose From	Directions for Next Move
4a.	Lower leaves opposite, upper leaves alternate; flowers white	*Plagiobothrys scouleri*
4b.	All leaves alternate; flowers yellow or white with a yellow center	Go to 5
5a.	Flowers yellow; nutlets wrinkled and warty	*Amsinckia menziesii*
5b.	Flowers white with a yellow center; nutlets minutely bumpy or smooth	*Cryptantha intermedia*

Adapted from Plants of the Pacific Northwest Coast (Lone Pine Publishing) by Jim Pojar and Andy MacKinnon.

To use a dichotomous key, follow these steps:

1. **Read the first pair of questions or statements and decide which of the two choices fits your plant.**

   In Table 14-2, the first choice asks you to choose based on the color and arrangement of the flowers.

2. **Follow the direction at the end of the option you choose.**

   If you were looking at a plant with blue flowers in coiled clusters, you'd choose option 1b. When you look at the end of option 1b, the direction is to go to 3.

3. **Skip over any statements that you're not directed to go to.**

   If you choose option 1b in Step 2, you'd go straight to the pair of options at 3, skipping over number 2 in Table 14-2 entirely.

4. **Read the next pair of questions or statements and decide again which choice best fits your plant.**

   The choices at 3 ask again about flower color and include some details about the fruit (a nutlet). If your flower is blue, you'd choose option 3a.

5. **Repeat these steps until you arrive at the name of your plant.**

   The choice of option 3a leads you to the name *Myosotis laxa,* which is the scientific name for your plant.

Chapter 15

Examining the Forest Floor: Bryophytes and Seedless Vascular Plants

. . . .

In This Chapter
▶ Exploring how plants moved from the ocean to the land
▶ Discovering bryophytes
▶ Examining seedless vascular plants

. . . .

Lots of people love flowering plants, and many admire the beauty of tall pines and redwoods. But have you ever stopped to admire the smaller members of the plant kingdom, such as the mosses and liverworts that help carpet the forest floor? Have you ever walked through the woods or bracken and admired the feathery ferns? This chapter looks at the structure and life cycles of these somewhat lesser known members of the plant kingdom, which have many features in common with their showier relatives.

Moving onto the Land

The closest living relatives of land plants are green algae called charophytes. The earliest plant fossils suggest that plants moved onto the land between 490 and 443 million years ago (during the Ordovician Period). Botanists think that the transition from life in water to life on land probably occurred in shallow ponds and marshes, where aquatic plants had access to water but also experienced higher levels of light. Plants that had characteristics that enabled them to move higher onto the land would benefit from the lack of competition for resources and could produce more offspring than the plants that had to stay totally submerged. Over long periods of time, a population of plants evolved to have features enabling it to survive on land.

The adaptations that enable plants to survive on land include both vegetative and reproductive adaptations:

- Vegetative adaptions include
  - Vascular tissue to transport water throughout the plant
  - A waxy cuticle that prevents plant tissues from drying out, but that contains stomates so that plants can still exchange gases with the atmosphere
  - Formation of a symbiotic association with soil fungi (mycorrhizae) that helped land plants acquire water and minerals from the soil
  - Protective pigments that help block ultraviolet radiation
- Reproductive adaptations include
  - Thick spore walls to protect reproductive spores from drying out
  - Tissues called gametangia that surround and protect gametes (eggs and sperm) from drying out
  - Embryos protected by maternal tissues

The first plants that colonized the land probably had all of these characteristics and were very similar to plants called bryophytes and seedless vascular plants (see Figure 15-1) that are alive today. Phylogenetic studies on the relationships between living plants support the idea that modern bryophytes are descended from the earliest land plants.

Chapter 15: Examining the Forest Floor: Bryophytes and Seedless Vascular Plants

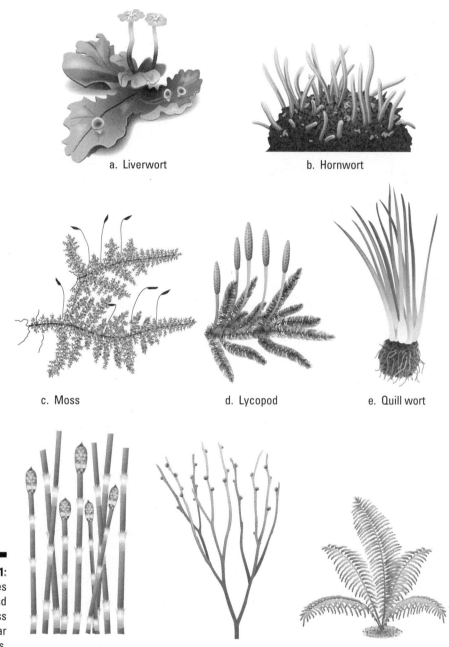

Figure 15-1: Bryophytes and seedless vascular plants.

a. Liverwort
b. Hornwort
c. Moss
d. Lycopod
e. Quill wort
f. Horsetail
g. Whisk fern
h. True fern

Bryophytes: Nonvascular Plants

Bryophytes are small plants that typically grow in moist areas. The bryophytes that are most familiar to you are probably mosses, which can be found growing in forests all over the world.

Bryophyte and bryophytes are informal names for a diverse group of nonvascular plants. These terms aren't the same as Bryophyta, which is a formal taxon (phylum) that refers specifically to the true mosses. (See the upcoming section "Mosses: Bryophyta.")

Bryophytes have several common characteristics:

- The gametophyte generation is the dominant, or most visible, generation.
- The structure of the plant body is simple and typically lacking vascular tissue like xylem and phloem.
- Bryophytes need water from rain or dew in order to reproduce.
- The water content of the plant body is about equal to that of the environment. When conditions are dry, the plants dry out, but don't necessarily die. When conditions are wet, the plants begin to grow again.

Organisms, such as bryophytes, that can survive severe drying out are called *poikilohydric*.

Bryophytes, which are organized into three phyla, all have certain characteristics in common:

- The gametophyte is the dominant phase of the life cycle.
- Structures called antheridia produce sperm, and archegonia produce eggs.
- They require water for reproduction so the sperm can swim to the egg.
- After fertilization, the zygote develops inside the archegonium, and the sporophyte remains dependent upon the gametophyte.
- They release their spores from elevated sporangia.

Liverworts: Phylum Hepatophyta

Liverworts can be flat, lobed plants like the one shown in Figure 15-1A, or they can be leafy and look more like mosses. Liverworts got their name because people in the Middle Ages thought the lobed liverworts looked like the liver from a person — wort is an old word for "plant," so liverwort literally means "liver plant."

Chapter 15: Examining the Forest Floor: Bryophytes and Seedless Vascular Plants

Structure of liverworts

The simple, flattened bodies of lobed liverworts are called *thalli* (singular: *thallus*). The thallus of a lobed liverwort is different on the top and the bottom (see Figure 15-2):

- The top surface has *pores* that are used for gas exchange. The pores remain open at all times.

- The lower surface has little root-like structures called *rhizoids* that are made from single cells. Rhizoids help anchor the liverworts to the ground. Some liverworts also have multicellular *scales* on their bottom surface.

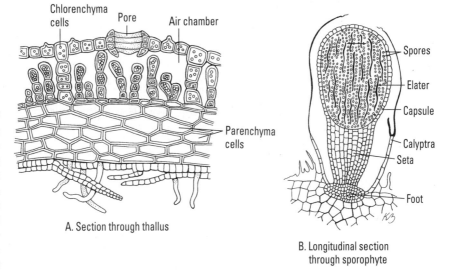

Figure 15-2: Cross-section through a complex, thallose liverwort (*Marchantia*).

A. Section through thallus

B. Longitudinal section through sporophyte

Some thallose liverworts, such as the liverwort *Marchantia* shown in Figure 15-2, have a fairly complex internal structure. The pores lead into airy pockets called *air chambers* that contain little stacks of photosynthetic chlorenchyma cells. Layers of nonphotosynthetic parenchyma cells form the bottom of the thallus. (For descriptions of different types of plant cells like chlorenchyma and parenchyma, check out Chapter 2.)

Life cycle of liverworts

Liverworts begin as haploid spores. When a liverwort spore lands in a moist environment, it begins to grow by mitosis, producing a filament of cells called a *protonema*. One cell, called the *apical cell*, continues to produce new cells by mitosis to form a haploid thallus, such as the ones shown in Figure 15-1A and 15-2.

Sexual reproduction occurs when the haploid thallus produces gametangia called antheridia and archegonia:

- *Antheridia* **produce sperm.** Antheridia are little chambers on the surface of the thallus that become filled with many sperm cells.

- *Archegonia* **produce eggs.** Archegonia are little vase-like structures that form on the surface of the thallus. In liverworts, each archegonium contains a single egg in the bottom of the "vase."

Some liverworts are *monoecious*, meaning they produce male and female reproductive structures on the same plant. Others are *dioecious*, producing male and female reproductive structures on separate plants.

Because the thallus of liverworts is already haploid, sperm and eggs are produced by mitosis, not meiosis.

When water is available, the sperm use their flagella to swim to the eggs of a different liverwort plant and fertilization occurs, producing a diploid zygote. The zygote divides by mitosis to produce an embryo that develops into the diploid sporophyte. Liverwort sporophytes are small club-shaped structures that grow out of the archegonia that they were formed in.

Cells within the sporophyte called *spore mother cells* divide by meiosis to produce haploid spores, completing the liverwort life cycle.

Liverworts can also reproduce asexually by *gemmae,* little groups of cells seen in cuplike structures called gemmae cups. The gemmae can be splashed out of the cup by rain drops and grow into a new plant.

Marchantia

Scientists and teachers often use the liverwort, *Marchantia*, as an example of liverwort structure and life cycle. In *Marchantia,* the antheridia and archegonia are lifted up on little stalks, producing umbrella like structures called *gametophores* (-phor means to carry, and these structures will carry gametes):

- The male gametophore is called the *antheridiophore*. The top of the antheridiophore in *Marchantia* looks like a flat disk (like the ones in Figure 15-1A).

- The female gametophore is called the *archegoniophore*. The top of the archegoniophore in *Marchantia* looks like the top of a palm tree (flip back to Chapter 11 for a drawing).

Marchantia is dioecious, so some plants produce only antheridiophores, while other plants produce only archegoniophores.

The sporophyte that develops after fertilization, which is drawn in Figure 15-2B, is designed to produce spores and then eject them over a wide area. The sporophyte has three main regions:

Chapter 15: Examining the Forest Floor: Bryophytes and Seedless Vascular Plants

- The foot attaches the sporophyte to the archegoniophore.
- The seta is a short thick stalk that suspends the rest of the sporophyte from the archegoniophore.
- The capsule is the main part of the sporophyte.

  The capsule contains different types of cells. The spore mother cells undergo meiosis to produce new spores. *Elaters* are long, pointed cells that change shape rapidly, helping to eject the spores over a wide area. The elaters change shape in response to changes in humidity, twisting and untwisting to help push spores out of the sporophyte. The *calyptra* is a layer of cells that protects the sporophyte until it is mature. When the sporophyte is mature, the calyptra breaks, and the capsule splits open, allowing the spores to leave the sporophyte.

Hornworts: Phylum Anthocerophyta

Hornworts are tiny horn-shaped plants, like the one drawn in Figure 15-1B, that typically grow in moist shaded areas. In tropical rainforests, they may cover large patches of soil or grow up onto the sides of trees.

The gametophytes of hornworts (Figure 15-1B) grow as flat, slimy-looking, blue-green patches. The twisty, horn-shaped parts of the plants are the sporophytes that grow up and out of the gametophytes.

A group of cells near the bottom of the sporophyte continues to divide throughout the life of the sporophyte so that the sporophyte keeps getting longer from growth at the base. This group of cells forms a *near-basal meristem*. Hornworts are the only plant group that has a meristem in this location.

Like liverworts (see previous section), the thalli of hornwort gametophytes have openings called pores that lead to chambers within the thallus. However, hornwort gametophytes have two distinct differences from liverwort gametophytes:

- The chambers within the thalli of hornworts are filled with a gelatinous substance called *mucilage*, which is why the gametophytes of the plants appear slimy.
- When you look at the cells of a hornwort under a microscope, each cell contains a single, large chloroplast rather than many chloroplasts, as is more typical in plant cells.

The sporophytes of hornworts have true stomata that can open and close and probably function in gas exchange.

Mosses: Phylum Bryophyta

Although people don't tend to notice them much, mosses are a very diverse group of plants — there are almost twice as many kinds of mosses as there are mammals.

Lots of small, leafy green stuff gets called moss, but it's not always a true moss. The pale green, plant-like material called reindeer moss is actually a lichen, the dark red leafy Irish moss is actually a red alga, and the long, thin, pale green strings called Spanish moss that hang from trees are actually flowering plants. To tell whether something is really a moss, look for the characteristics that I describe in this section.

Mosses can be found all **around** the world, growing on the forest floor, along the banks of rivers and streams, and on the arctic permafrost. They're important in the environment because they help with water and nutrient control, help prevent erosion, and help insulate the permafrost.

Structure of mosses

Moss gametophytes, like the one drawn in Figure 15-1C, usually look leafier than those of other bryophytes. The leaves of mosses are usually very simple and just one cell thick, although some moss leaves have a *midrib,* a thicker area that runs down the center of the leaf.

Moss gametophytes differ from those of other bryophytes in several ways:

- Moss gametophytes have *radial symmetry,* with their leaves arranged around a central axis, whereas the gametophytes of both liverworts and hornworts are flattened and have bilateral symmetry.
- The rhizoids that anchor moss gametophytes to the ground are multicellular, whereas the rhizoids of liverworts are unicellular.
- The leaves of mosses tend to be pointier and more leafy-looking than the leaves of leafy liverworts, which are more rounded and scale-like.

Although bryophytes in general don't have vascular tissue, some mosses have cells that are very similar to vessels and sieve cells.

Moss sporophytes look like skinny little lanterns or flags growing up off the gametophyte (you can see some in Figure 15-1C). The sporophyte has several parts, which are shown in Figure 15-3:

- The sporophyte grows upward on a stalk called a seta.
- A chamber called a capsule forms at the end of the seta.
- The capsule is covered by a little hat-like piece of tissue called a calyptra.

Chapter 15: Examining the Forest Floor: Bryophytes and Seedless Vascular Plants

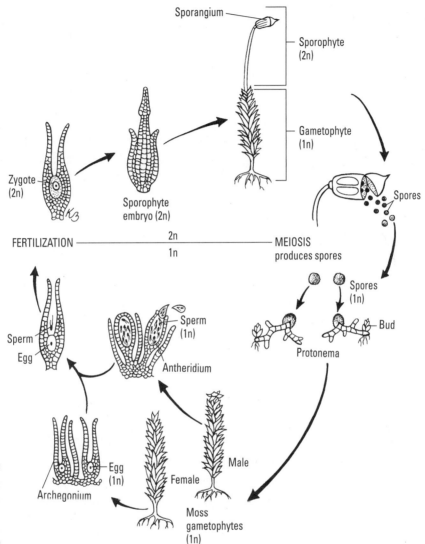

Figure 15-3: The life cycle of a moss.

Life cycle of mosses

Beginning with the germination of a haploid spore, the moss life cycle has the following steps (refer to Figure 15-4):

1. **The spore germinates, growing by mitosis into a filament of cells called a *protonema*.**

   As the protonema grows, it branches and forms a mat-like network.

2. **At various points along the protonema, cells divide by mitosis to produce small leafy structures that develop rhizoids and grow into the leafy gametophytes.**

3. **At the tips of the gametophytes, or sometimes on special branches, cells differentiate to form gametangia.** A cell within each vase-shaped *archegonium* divides by mitosis to produce an egg. Cells within the sac-like *antheridia* divide by mitosis to produce many sperm. As in other bryophytes, some mosses are monoecious, while others are dioecious.

   Sterile hairs called *paraphyses* grow between several archegonia or between several antheridia on different stems of a single plant.

4. **When water is available, the sperm swim to the egg, attracted by substances released by the archegonia.**

   The sperm fertilizes the egg, forming the diploid zygote.

5. **The zygote grows by mitosis into an embryo, which breaks down cells at the base of the archegonium and then attaches itself with a knob-like mass of cells called a *foot*.**

6. **The sporophyte grows up and out of the archegonium, taking the top of the archegonium with it, which forms the little hat-like calyptra.**

   As the sporophyte grows, it develops chloroplasts, becomes green, and starts to do photosynthesis (see Chapter 7). However, the sporophyte continues to receive some carbohydrates, as well as water and minerals, from the gametophyte by absorbing them through its foot.

7. **The mature sporophyte, which consists of the capsule (also called a sporangium) supported on a seta, produces spores.**

   Cells within the sporangium, called *spore mother cells,* divide by meiosis to produce the haploid spores.

8. **When the spores are mature, the lid of the capsule, called the *operculum*, falls off, revealing an opening called the *peristome*.**

   Moss spores are carried away by the wind once the operculum falls off. In many mosses, the peristome is ringed with a row of triangular, membranous teeth. These teeth change shape in response to moisture, opening and closing the peristome.

Asexual reproduction of mosses can also occur through *fragmentation,* when pieces of moss gametophyte break off and then regrow to form new individuals.

> ### For peat's sake!
>
> Peat moss is made of a moss called *Sphagnum* that has some very unique and useful characteristics. If you've ever used peat moss in gardening, you probably know that it is good at holding water in the soil — but did you know that it can hold up to 25 times its own weight in water? Peat moss is so great at holding water that Native Americans used it as baby diapers. And yet, when it's dry, peat makes great fuel — people that live near peat bogs have been cutting blocks of peat and using them for fuel for centuries.
>
> Another characteristic of *Sphagnum* is that it produces acids as it grows, making its environment acidic. The acid prevents bacteria from growing, so peat bogs are places where decomposition of dead materials happens very slowly. Perfectly preserved bodies have been found in peat bogs, like the Tollund man that was found in the 1950s. Scientists estimated that the man lived about 400 years BC, yet when he was found his face was clearly visible, and his clothes and hair were intact — as was the rope around his neck that was used to hang him almost 2,500 years ago!
>
> One less creepy result of the fact that bacteria don't grow in the acid is that peat makes great bandages. And not only is it antimicrobial, it's more absorbent than cotton. In fact, when cotton bandages ran low during World War I, peat had a successful run as a replacement bandaging material.

Seedless Vascular Plants

Once plants like bryophytes became adapted to living on the land, evolution continued to select for plants that had traits favorable to this strange, dry territory. Today's descendants of these early plants reveal the traits that made them successful:

- **Better control of internal water concentration.** Plants developed thicker cuticles, providing better waterproofing. Plants also developed stomata that could open and close became the norm, rather than pores like those in bryophytes that stay open all the time.

- **Development of roots to increase water and mineral uptake from the soil.** Scientists think roots evolved in ancient vascular plants from stems that grew along or just under the surface of the soil. Once these surface branches started growing under the surface of the soil, they were exposed to different conditions than the branches above ground. Plants whose underground branches were able to absorb water had an advantage over other plants, leading to the evolution of a more root-like structure over time.

- **Internal vascular tissue for conducting water and sugar throughout the plant body.** In particular, the development of water-conducting tracheids helped change the way plants grow. Because of their strong, lignified cell walls, tracheids provide greater structure and support to plants, enabling plants to grow more erect.

- **A more erect, branched pattern of growth.** Branching let the early land plants produce multiple sporangia per plant, increasing their chances of reproductive success. And once plants started producing leaves, branching allowed plants to spread their leaves out so that they didn't shade each other, maximizing the capture of sunlight for photosynthesis.

- **The evolution of leaves, allowing better capture of light.** Small, simple leaves, called *microphylls*, evolved in a group of plants called lycophytes. Microphylls are simple leaves with a single unbranched vein of vascular tissue. In addition, larger, more complex leaves called *megaphylls* developed in almost all other vascular plants. Megaphylls have branching veins. Scientists think megaphylls evolved when leaf tissue grew around small, flat clusters of branches, joining them together in a leaf-like structure. Over time, the branches became the branching veins within the leaf.

The combination of these new traits gave plants much better control over their internal water content, increased their reproduction, and improved their ability to effectively harvest light for photosynthesis.

Organisms that can regulate their internal water content are called *homiohydric*.

Another important characteristic of plants that first appears in the seedless vascular plants is heterospory. Heterosporous plants produce two different types of spores:

- *Megaspores* are larger and grow into female gametophytes.
- *Microspores* are smaller and grow into male gametophytes.

Within the plant kingdom, heterospory evolved many times, so scientists think it must have given plants a real advantage. Many plant groups today, including the very successful flowering plants, are heterosporous. Plants whose sporophytes make just one kind of spore are *homosporous* (*homo-* means same).

After these evolutionary innovations occurred, groups of land plants expanded and evolved into many different species, creating a burst in plant diversity between 417 and 354 million years ago (the Devonian Period) that eventually led to the development of the first forests between 354 and 290 million years ago (the Carboniferous period).

Over these time periods, the relationship between the gametophyte and sporophyte generation changed, and sporophytes became the more dominant plant generation with gametophytes getting progressively smaller and more protected by the sporophyte. Botanists think that the reason sporophytes became dominant is that all the adaptations that allow plants to survive in dry environments evolved in sporophytes.

All seedless vascular plants have several characteristics in common:

- The sporophyte is the dominant phase of the life cycle, but the gametophyte is independent and photosynthetic.
- The sporophyte makes and disperses spores from sporangia.
- The gametophyte makes gametes in gametangia.
- They require water for sexual reproduction so the sperm can swim to the egg.
- The sporophyte has vascular tissue, a cuticle, and stomates.

Club mosses: Phylum Lycophyta

The lycophytes that are alive today are small plants that look a lot like mosses (see Figure 15-1D). These looks are deceiving, however, as lycophytes are definitely not mosses.

In mosses, the gametophytes are the more visible part of the life cycle, but in lycophytes, the sporophytes are more visible. Lycophyte gametophytes are very tiny and some even live underground, relying upon a symbiotic relationship with a fungus to provide them with nutrients and water.

Many lycophytes grow from rhizomes that branch *dichotomously,* periodically splitting into two forks. New shoots grow off the rhizome and may stand upright or may dangle off to the side of the rhizome.

Three groups of lycophytes exist today:

- **Lycopods,** which are often called club mosses or ground pines can be found looking a lot like little pine trees as they grow along the forest floor in temperate regions. (The drawing in Figure 15-1D is a lycopod.)
- **Spike mosses** are flatter and more moss-like in their appearance than the lycopods. Some species of the spike moss *Selaginella* are known for their ability to survive in harsh, dry conditions, such as rocky desert outcrops. Other species grow within the tropical rainforest canopy.
- **Quillworts** are spiky little plants that almost look like tufts of grass and are typically found growing in marshy areas.

Structure of lycophytes

Lycophytes have true leaves, roots, and stems that contain vascular tissue. The leaves of lycophytes are microphylls.

The sporophytes of each group of lycophyte have a unique growth form, which are compared in Table 15-1:

- **Club mosses:** The branching rhizomes of club mosses, such as those in the genus *Lycopodium*, grow along the forest floor, periodically growing little shoots that either stand up like little pine trees or dangle off the rhizomes (see Figure 15-1D). Most lycopods are less than a foot tall, but some tropical species grow up to five feet long.
- **Spike mosses:** The rhizomes of spike mosses, such as those in the genus *Selaginella*, are more branched than those of lycopods.
- **Quillworts:** Quillworts, such as the one drawn in Figure 15-1E, belong to the genus *Isoetes*. They grow in dense little clusters and on the surface look quite different from other lycophytes.

Table 15-1 A Comparison of Lycophyte Groups

	Club mosses	*Spike mosses*	*Quillworts*
Branches	Shoots may be branched or unbranched. Shoots that do branch split evenly, making symmetrical forks in a type of branching called *isotomous branching*.	Shoots usually look flat, and their branching is typically *anisotomous* — when the shoot splits, one branch at the fork grows longer than the other.	The short, thick stems of quillworts form perennial undergound corms.
Leaves	Small leaves wrap around the stems, either in a spiral or in individual rings of leaves called whorls.	Small leaves are arranged in spirals around the shoot. At the base of each leaf is a small flap of extra tissue called a *ligule* (ligule means tongue).	Longer leaves are arranged in a tight spiral around the short stem, growing up like tufts of quills. Bases of leaves have ligules
Roots	Adventitious	Adventitious roots, sometimes called *rhizophores*, look like little brown strings dangling down from the stem and then growing into the soil.	Adventitious and dichotomously branched

Here there be giants

Although lycophytes today are mostly small plants, at one time giant lycophytes ruled the earth. During the Carboniferous period (354 to 290 million years ago), lycophyte trees over 100 feet tall grew in hot and humid swamps along with other types of plants. These giant lycophytes, such as the tree *Lepidodendron*, looked very similar to their smaller relatives, with dichotomously branching roots and shoots. The leaves fell off the upward growing trunks of the trees, leaving very distinctive diamond shaped patterns of leaf scars. Because the scars look like scales, these trees are sometimes referred to as scale trees. Climate change at the end of the Carboniferous led to the death of the plants and caused their remains to be covered with seawater. The huge masses of plant material were buried in such a way that decomposition was very slow, allowing the carbon-containing plant material to be transformed into the coal and other fossil fuels that humans rely upon today.

Life cycles of lycophytes

Within the lycophytes, club mosses are homosporous, while spike mosses and quillworts are both heterosporous.

When lycophyte sporophytes are ready to make spores, they produce little bean-shaped sporangia that are tucked into the pockets, or *axils*, where the leaves meet the stem. (Check out Figure 15-4 to see a drawing of the sporangia tucked into the leaves.)

Leaves that have sporangia on them are called *sporophylls* (*spor* refers to the spores, *phyll* means leaf, so sporophylls are spore-bearing leaves). In heterosporous lycophytes, leaves that bear *megasporangia* are called *megasporophylls*, while leaves that bear *microsporangia* are called *microsporophylls*.

In some lycophytes, the sporophylls look very similar to other leaves and are mixed in among them in zones along the stem. In others, the sporophylls are small and brown and clustered together in cone-like structures called *strobili*. You can see strobili on the club mosses in Figure 15-1D and 15-4.

Figure 15-4 shows the life cycle of the club moss, *Lycopodium*, as an example of the life cycle of homosporous lycophytes:

1. **Mature diploid sporophytes produce sporangia on sporophylls.**

   Within the sporangia, spore mother cells divide by meiosis to produce haploid spores.

2. **The spores are released into the air, and when they land in a suitable environment, they begin to grow by mitosis to produce the haploid gametophyte.**

3. **Club moss gametophytes are monoecious, so each gametophyte produces both archegonia and antheridia on its surface.**

    Cells within archegonia divide by mitosis to produce a single egg in each archegonium. Cells within antheridia divide by mitosis to produce sperm.

4. **When moisture is available, the sperm swim to the eggs and fertilize them, forming the diploid zygote.**

    The zygote divides by mitosis, producing first the embryo and then growing into the sporophyte.

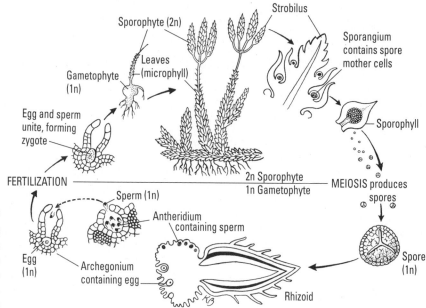

Figure 15-4: The life cycle of the club moss *Lycopodium*.

Some club mosses also reproduce asexually by producing little structures called *bulbils*, in the axils of their leaves. When bulbils fall off the plant, they can grow into new sporophytes.

Although the life cycle of the heterosporous lycophytes is very similar to that of homosporous lycophytes, some key differences exist. Figure 15-5 shows the life cycle of the spike moss, *Selaginella*, as an example:

1. **Mature diploid sporophytes produce sporangia on sporophylls.**

    Megasporangia are formed on megasporophylls, while microsporangia are formed on microsporophylls.

2. **Within the sporangia, cells divide by meiosis to produce haploid spores.**

Chapter 15: Examining the Forest Floor: Bryophytes and Seedless Vascular Plants

Inside megasporangia, megaspore mother cells divide by meiosis to produce megaspores. Inside microsporangia, microspore mother cells divide by meiosis to produce microspores. As the names suggest, megaspores are bigger than microspores.

3. **The sporangia release the spores into the air, and when the spores find a suitable environment, they grow by mitosis into gametophytes.**

   Megaspores grow into female gametophytes that will eventually make only archegonia. Microspores grow into male gametophytes that will make only antheridia. Because the two sexes are on separate plants, heterosporous lycophytes are dioecious.

4. **Cells within the gametangia divide by mitosis to produce gametes.**

   Cells within the archegonia divide to produce eggs, while cells within antheridia divide to produce sperm.

5. **When moisture is available, sperm swim to the eggs and fertilize them, forming the diploid zygote.**

   The zygote divides by mitosis to produce the embryo, and eventually the mature sporophyte.

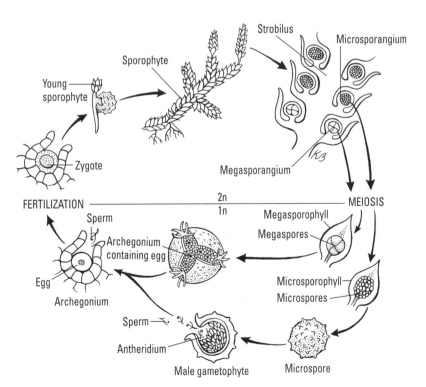

Figure 15-5: The life cycle of the spike moss, *Selaginella*.

Ferns and (some of) their allies: Phylum Pterophyta

Scientists sometimes use the phrase fern allies to refer to the seedless vascular plants like the lycophytes, horsetails (see Figure 15-1F), and whisk ferns (see Figure 15-1G). Until recently, scientists considered the ferns and their allies to be four separate phyla: Lycophytes were grouped in the Lycophyta, horsetails in the Sphenophyta (or Equisetophyta), whisk ferns in the Psilophyta, and true ferns in the Pterophyta (or Pteridophyta or Polypodiophyta). Based on comparisons of their DNA sequences, however, scientists now feel that although the lycophytes belong in their own phylum, the other fern allies are more closely related to ferns and actually belong with them in the same clade. So, some scientists have reorganized the Pterophyta (or monilophytes) to include three groups, which are compared in Table 15-2:

- **Horsetails and scouring rushes,** which are grouped in the genus *Equisetum,* literally grow like weeds all over the world. If you've ever pulled one out of your yard, you know firsthand that they're rough and scratchy to the touch. The reason they're so scratchy is because the plants deposit silica — the same element that makes up sand — in the walls of the epidermal cells that make up the stem. The silica gives the plant a gritty feel, which hasn't gone unnoticed by people who have a long history of using these plants as tools for scouring and even sharpening objects.

- **Whisk ferns,** which are in the genus *Psilotum,* got their name from their resemblance to little whisk brooms: they have no leaves or roots, so they're just clusters of branching stems. Their unique appearance resembles that of an extinct group of early plants called *psilophytes*, and for a long time scientists thought they might be descended from that group of ancient plants. The structure of their gametophytes, however, is very similar to that of true ferns, and recent comparisons between the DNA of these two groups confirm that they are close relatives. Now scientists think that whisk ferns are descended from true ferns that evolved to stop making leaves.

- **True ferns** have relatively large, often very beautiful leaves called *fronds*, so that many people have them as houseplants. Ferns range in size from tiny floating ferns about a half-inch in size to tall, tropical tree-ferns that can grow over 75 feet tall. If you want to take a walk in a forest that will make you feel like you've gone back to the Carboniferous Period, you can visit forests of tree ferns in Australia and New Zealand.

The leaves of pterophytes are megaphylls.

Table 15-2 A Comparison of Ferns and Their Allies

	Horsetails	Whisk ferns	True ferns
Shoots	Perennial rhizomes grow underneath the soil, putting up shoots each year. Some shoots are annual; others are perennial. The stems are ribbed. Some horsetails are *dimorphic,* producing two kinds of shoots: green vegetative shoots that aren't involved in sexual reproduction and brown fertile shoots that produce spores.	Stems branch dichotomously, forming an isotomous pattern (see this pattern in Figure 15-1G). Photosynthesis occurs in the outer cells of the stem.	Grow horizontally as rhizomes that periodically send up vertical leaves. Some ferns make short vertical rhizomes that produce many leaves over time, eventually becoming surrounded by old leaf bases. These stump-like structures can get quite large, especially on tree ferns.
Leaves	Tiny, but are evolved from megaphylls. At the nodes where the leaves attach to the stem, they fuse together forming a solid sheath of tissue.	Absent. Little green leaf-like flaps of tissue called *enations* occur at intervals along the stem. Enations are arranged in spirals around the stem, but they lack vascular tissue.	Initially form as tightly coiled structures called *fiddleheads* (or croziers), which are basically tightly rolled up fronds. As the fronds grow, the fiddle heads unroll and the frond expands. Many fern fronds, such as the one in Figure 15-1H, are pinnately compound, divided into leaflets called *pinnae* (singular: pinna). (For more on the structure of compound leaves, see Chapter 4).
Roots	Adventitious roots grow downward from the nodes along the stem.	Absent. Whisk ferns get water through rhizomes that are partnered with mycorrhizal fungi.	Adventitious roots along the rhizome.

Life cycles of pterophytes

Most pterophytes are homosporous, producing only one type of spore, which grows into a gametophyte that produces both antheridia and archegonia.

Figure 15-6 shows the life cycle of the horsetail *Equisetum* as an example of a pterophyte life cycle:

1. **The diploid sporophytes form cone-shaped reproductive structures called *strobili*, which support the sporangia that produce the spores.**

   In dimorphic horsetails, such as the one in Figure 15-6, the strobili form at the tips of special reproductive stems. In horsetails that aren't dimorphic, strobili are produced at the tips of photosynthetic stems. Small hexagon-shaped sporangiophores grow out sideways from the strobili, forming sporangia beneath each hexagon.

2. **Inside the sporangia, diploid spore mother cells divide by meiosis to produce haploid spores.**

   The spores of horsetails are wrapped with ribbon-like elators that extend in response to changes in humidity, acting like wings to help the spores fly through the air.

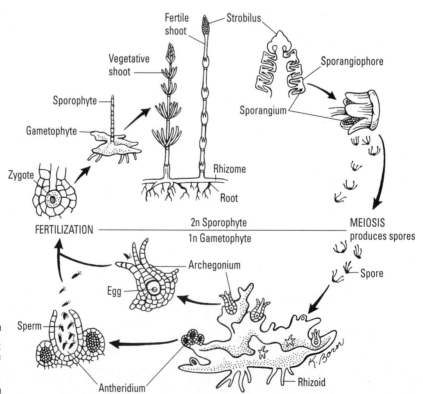

Figure 15-6: Life cycle of a horsetail.

3. **When the spores land in a good environment, they grow into tiny gametophytes.**

   At first, about half the gametophytes make archegonia, while half make antheridia:

   a. Archegonia develop along the surface of half of the gametophytes, and cells inside each archegonium produce an egg by mitosis. But after a while, these female gametophytes start producing antheridia as well. Scientists call this *protogynous* reproduction, meaning the gametophytes start out strictly female but then become bisexual. Protogyny helps prevent self-fertilization.

   b. Antheridia form immediately in about half the gametophytes, and cells inside the antheridia produce multiple sperm by mitosis. When water is available, changes in water pressure cause the sperm to explode out of antheridia so they can swim to the archegonia and fertilize the eggs.

4. **Fertilization forms a diploid zygote, which grows by mitosis into the embryo and then the mature sporophyte.**

Horsetails can also reproduce asexually. When the stems or rhizomes are broken, each fragment can grow into a new sporophyte.

Chapter 11 presents the fern life cycle as an example of plant life cycle with a dominant sporophyte generation. The basic life cycle is very similar to that of horsetails, with the biggest differences occurring in the sporangia and the gametophytes:

- The sporophytes form sporangia on the underside of the leaf.

  Often, the sporangia are clustered in little spots called sori (see Figure 15-7) that look rusty colored when mature. Different species of ferns have different patterns of sori, such as spots, stripes, and crescents. When the sporangia are immature, they may be covered by a protective flap of tissue called an *indusium*. As the sporangia mature, the indusia shrivel and peel back, exposing the sporangia.

  If you magnify a sorus, you would see the individual sporangia, which look a little like a rattle for a baby. The striped ridge of tissue that runs around the edge of each sporangium is called an *annulus*. The cells of the annulus have cell wall thickenings and are very sensitive to changes in moisture. When the spores are mature, the annulus snaps back, tearing open the sporangium and flinging the spores into the air.

- Fern spores germinate and grow into bisexual gametophytes that are tiny, delicate heart-shaped structures growing in moist environments. The gametophytes are only one cell thick in most places. Rhizoids form near the point of the heart on the lower surface and anchor the gametophyte to the ground.

- Flask-shaped archegonia develop near the notch in the heart shaped gametophytes. Cells inside each archegonium divide by mitosis to form an egg.

- Spherical antheridia develop nearer to the rhizoids at the point of the heart. Cells inside each antheridium divide by mitosis forming many sperm. When moisture is available, sperm swim to find to eggs to fertilize.

- Many ferns have ways to discourage self-fertilization. Some fern gametophytes develop their archegonia and antheridia at different times so their sperm won't fertilize their own eggs. Other fern gametophytes release hormones so that neighboring gametophytes are stimulated to produce gametes of the opposite sex.

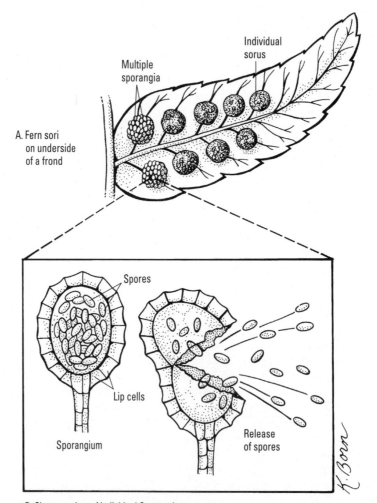

Figure 15-7: Release of spores from a fern sporangium.

Chapter 16

Their Seeds Are Naked: Gymnosperms

In This Chapter

▶ Examining reproduction in seed plants

▶ Exploring the diversity of gymnosperms

▶ Discovering the life cycle of pine

*T*he evolution of seeds gave a huge advantage to seed-bearing plants, helping them to diversify and spread over the land. One group of seed plants, the gymnosperms, produces its seeds on leaf-like structures that are not fully enclosed, leading botanists to refer to them as "naked seeds." Gymnosperms include very familiar plants like pine, spruce, and fir trees, as well as some plants you might never have heard of like cycads, *Gnetum*, and *Welwitschia*. This chapter explores the diversity of gymnosperms while also presenting the characteristics that all gymnosperms have in common.

Protecting the Embryo with Seeds

Most people probably take for granted that plants make seeds — after all, you can't eat a piece of fruit or cut open a tomato without seeing them. That's because seeds were such a good invention, that plants that make them became very successful and spread over the face of the earth.

Seeds are multicellular, complex structures that protect and nourish the embryo. Seeds can survive for long periods of time in the environment, waiting for conditions to be right for the growth of the embryo. (For more on seeds, flip to Chapter 5.)

The earliest evidence for seed plants comes from fern-like plants that grew to the size of trees during the late Devonian period, over 350 million years ago. At first, botanists grouped these plants with the true ferns, but then they noticed that the fronds had seeds on them, not spores. (For details on spores and true ferns, see Chapter 15.) So, botanists put these plants into their own category, called *pteridosperms,* which means seed ferns. Pteridosperms are classified in the plant group called gymnosperms.

Gymnosperms produce seeds and pollen and have vascular tissue, but they don't make flowers or fruits.

The word gymnosperm literally means naked seeds (*gymnos*=naked, *sperma*=seed) because the seeds of gymnosperms aren't fully enclosed inside of a structure. The other group of seed-bearing plants, the flowering plants or *angiosperms,* enclose their seeds inside of fruits. Figure 16-1 compares the seed-bearing structures of gymnosperms with those of angiosperms.

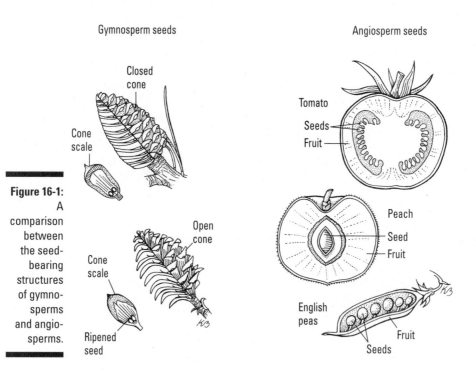

Figure 16-1: A comparison between the seed-bearing structures of gymnosperms and angiosperms.

Although gymnosperms and angiosperms package their seeds differently, both groups developed similar reproductive strategies:

- **Gymnosperms and angiosperms are both *heterosporous*.** Sporophytes produce two different types of spores that give rise to unisexual gametophytes. (For a description of heterospory, see Chapter 15.)

    - Megasporangia produce megaspores that grow into female gametophytes.
    - Microsporangia produce microspores that grow into male gametophytes.

- **Gametophytes are very small and completely enclosed within ovules and pollen grains.** The outer sporophyte tissues of the ovules and pollen grains protect the fragile gametophytes inside, keeping them moist and shielding them from ultraviolet radiation.

    The male and female gametophytes are each protected by a unique structure:

    - **The male gametophyte is protected by the *pollen grain*.** Microspores develop into pollen grains that are covered with a protective coat. When pollen grains land on the female part of a seed plant, they begin to grow a long tube called a *pollen tube* that delivers sperm to the egg inside the female gametophyte

    - **The female gametophyte is protected by the *ovule*.** The megasporangium and megaspore are wrapped with layers of sporophyte tissue called *integuments,* forming a structure called an *ovule*. Inside the ovule, the female gametophyte produces one or two egg cells. After fertilization, ovules develop into seeds.

Figure 16-2 shows examples from the four groups of gymnosperms that are alive today:

- **Cycads,** like the one in Figure 16-2A, look a little bit like short palm trees because they have a crown of spiky, compound leaves that emerge from their stocky trunk.

- **Gnetophytes** are unusual plants that include *Gnetum,* shown in Figure 16-2B, and the large-leaved *Welwitschia,* shown in Figure 16-2C.

- **Ginkgos,** like the one in Figure 16-2D, may be familiar to you because they're often planted along city streets. Their delicate, fan-shaped leaves are also often seen in Asian art.

- **Conifers** are probably very familiar to you as this group includes pines, cedars, spruces, firs (Figure 16-2E), redwoods (Figure 16-2F), and hemlocks.

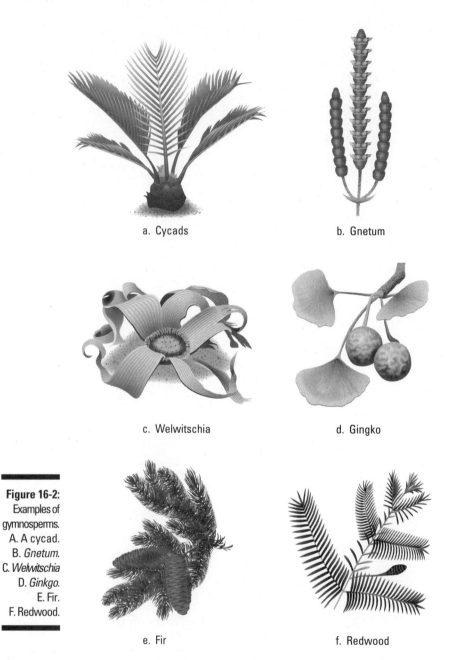

Figure 16-2: Examples of gymnosperms. A. A cycad. B. *Gnetum*. C. *Welwitschia* D. *Ginkgo*. E. Fir. F. Redwood.

a. Cycads
b. Gnetum
c. Welwitschia
d. Gingko
e. Fir
f. Redwood

Cycads: Cycadophyta

Very few cycads remain living in the wild today, but during the Jurassic period (206 to 144 million years ago), they dominated the landscape in many parts of the world. Today, cycads still grow wild in some tropical and subtropical areas and are also seen as ornamental plants. Ornamental cycads are sometimes referred to as palms, but they aren't really closely related to palm trees, which are flowering plants.

Cycads have several interesting characteristics:

- **The cycad trunk is unbranched.** If the trunk is damaged, sometimes branching will occur.
- **Cycads produce some roots that grow upward and form symbiotic partnerships with blue-green bacteria.** Blue-green bacteria can capture nitrogen from the atmosphere and convert it to forms usable by plants. For more on this process, called *nitrogen fixation,* see Chapter 19.
- **Insects pollinate cycads.** Beetles are attracted to the scent and heat of the pollen. (Cycads may have evolved insect pollination 100 million years before the flowering plants!)
- **Cycads are *dioecious:* Female plants produce only female cones, and male plants produce only male cones.** Female cycad cones can be quite large, up to 2 feet long and weighing 90 pounds!
- **Cycad sperm have flagella and can swim.** Cycads and ginkgos are the only seed plants that have swimming sperm.
- **Animals disperse the large, brightly colored seeds.** In Africa, elephants eat the whole cone of some cycads and then disperse the seeds in their dung.

Unfortunately, cycads grow very slowly, and their numbers are dwindling due to habitat destruction and overharvesting by humans.

Ginkgoes: Ginkgophyta

Only one species of ginkgo, *Ginkgo biloba* or the maidenhair tree, is still alive today. Because ginkgo trees are pretty as well as resistant to pests and pollution, they're often planted as ornamental trees in cities.

Although ginkgoes look very different from other gymnosperms, they have other characteristics like naked seeds that place them firmly in this group:

- **Ginkgo leaves are flattened and *deciduous,* with fine, dichotomously branching veins.** In autumn, the leaves turn a beautiful yellow color before they fall from the trees.
- **Like cycads, ginkgoes are dioecious: Male and female parts occur on different trees.** The seeds of ginkgo have a strong, unpleasant odor, so typically only males are planted along city streets.

- **Also like cycads, ginkgos have motile sperm.** The sperm swim to the egg to achieve fertilization.
- **Gingko seeds have an outer, fleshy covering.** Although the covering has a rancid odor, the inner part of the seed has nutritious and medicinal properties. Certain cultural groups in China have a long history of cultivating ginkgos for their seeds.

Conifers: Coniferophyta (Pinophyta)

Conifers are the largest and most diverse group of living gymnosperms, second only to the angiosperms in terms of distribution of seed plants around the world. Members of this group range in size from juniper shrubs to giant redwood trees, and some form large forests in areas of the Northern Hemisphere that have short growing seasons.

Conifers get their name because they typically produce their seeds in woody *cones* (*conus*=cone and *ferre*=to carry), so conifers are literally cone-bearing plants. However, some types of conifers, like junipers and yews, produce unusual cones or even no seed cones at all:

- **The female cones of junipers are fleshy.** These round, bluish cones are often called *juniper berries,* but they're not berries at all.
- **Yew trees don't produce female cones.** The seeds of yew are surrounded by a red, fleshy layer called an *aril,* which gives the seeds a berry-like appearance. Not only aren't arils really berries, but their seeds are poisonous as well.

Except for differences in cones, conifers share certain characteristics:

- **Conifers have simple leaves with single veins.** Although the leaves of conifers are very small, they aren't microphylls. Fossils of extinct conifers show that the ancestors of modern conifers had megaphylls, so the small leaves of modern conifers must have evolved as a reduction in size from the original megaphylls. (For a comparison of microphylls and megaphylls, see Chapter 15.)
- **Male and female gametophytes are produced on separate male and female cones.** Unlike cycads and ginkgoes, however, many conifers are monoecious, producing both male and female cones on the same plant.
- **Conifers are wind pollinated.** The wind carries pollen from male cones to the female cones. Once the pollen reaches the ovules, it grows a pollen tube and delivers sperm to the egg.
- **Wind and animal dispersal of seeds occurs.** Some wind-dispersed seeds have flattened wing-like structures that help them catch flight. Fleshy cones and structures attract animals like birds and squirrels, who eat the cones and scatter the seeds in their feces. Some pine seeds, like the pine "nuts" of the pinyon pine, are prized by people and other animals.

A cancer-fighting tree

The Pacific yew tree, *Taxus brevifolia,* has saved many lives. Pacific yews grew quietly in the forests of the Pacific Northwest, and people considered them of little value next to the taller Douglas Fir and Sitka Spruce. But then, in the 1960s, the USDA went looking for anticancer drugs, starting a huge screening program that examined thousands of plants for any chemicals that were active against tumors. The bark of the Pacific yew yielded a chemical, later named *taxol,* that showed anticancer activity. Taxol stops cancer cells from multiplying and is used to treat breast and ovarian cancer, as well as Kaposi's sarcoma. For a while, taxol was in such high demand that the survival of the Pacific yew was threatened by overharvesting of its bark. Fortunately for the trees, scientists developed a synthetic version of this valuable drug.

Pines

Pine trees belong to the largest genus of conifers, the genus *Pinus.* Pine trees dominate the coniferous forests of the Northern Hemisphere. The xylem of pine trees contains tracheids, but no vessels or fibers, making it a *soft wood* tree.

Structure of pines

Pine trees often grow in cold, dry, or windy areas where protecting themselves from water loss is a serious issue. The cross-section of a pine needle in Figure 16-3 shows some features of pine needles that help them resist water loss and pest damage:

- The epidermis has a thick, waxy cuticle.
- The stomata are sunk into the surface of the leaf, helping to reduce the rate of transpiration (see Chapter 9).
- Special resin-producing cells line channels called *resin canals,* producing resin that protects the needles from fungi and insects.
- The vascular tissue in the leaf is surrounded by a layer of endodermis. In most other plants, endodermis is found only in the roots (see Chapter 4).

A distinctive characteristic of pines is that their needle-like leaves are clustered together in bunches of between two to five needles. The needles in each bunch, or *fascicle,* fit together to form a cylindrical shape.

Life cycle of pine

All you ever really see of a pine tree (or any gymnosperm) is the sporophyte generation because the gametophyte generation is very small and enclosed within sporophyte tissue.

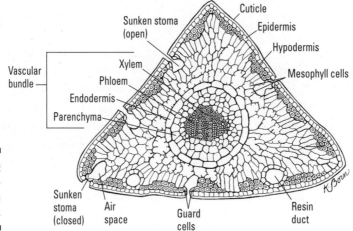

Figure 16-3: A cross-section of a pine needle.

Figure 16-4 illustrates the life cycle of a pine tree, showing how the sporophyte and gametophyte generations alternate:

1. **Diploid sporophyte trees produce both male and female cones.**

    Male cones, or strobili, are smaller and made up of papery scales, while female cones are larger and made up of woody scales.

2. **Within the cones, sporangia develop at the base of the scales.**

    Two microsporangia form at the base of each scale in male cones.

    A pair of ovules, each containing a megasporangium, develops at the base of the scales in female cones. Each pine ovule is enclosed in layers of protective sporophyte tissue called integuments.

3. **Within the sporangia, mother cells, or meiocytes, undergo meiosis to produce haploid spores** (for details on meiosis, see Chapter 11).

    Microspore mother cells produce four haploid microspores that develop into pollen grains. Each pollen grain consists of four cells and a pair of air sacs that stick off the pollen grain like mouse ears. The air sacs make pine pollen light and easily carried long distances by the wind.

    Megaspore mother cells produce four megaspores. Only one megaspore survives, growing into the multicellular female gametophyte. The female gametophyte develops slowly, sometimes taking months before it's mature. Toward the end of its development, the gametophyte produces two to six archegonia, each of which contains a single egg.

4. **Pollination occurs when the wind carries pine pollen from the male cones to the female cones.**

    Pollen enters the ovules through a small hole between the integuments called the *micropyle.* The pollen grain germinates, or begins to grow, producing a pollen tube that grows toward the female gametophyte.

5. **Fertilization occurs when a sperm cell fuses with an egg cell.**

Chapter 16: Their Seeds Are Naked: Gymnosperms

When the pollen tube reaches the female gametophyte, it releases two sperm cells. One sperm cell fertilizes the egg; the other breaks down.

6. **After fertilization, the diploid embryo develops, and the ovule matures into a seed.**

   The embryo, which represents a new sporophyte generation, is protected and nourished by gametophyte tissue inside the seed. The integuments that surrounded the ovule harden and become the seed coat. It takes 15 months from the time of fertilization for the seed to fully mature!

The pine seed contains tissue from three different generations: the seed coat that developed from the integuments of the parent sporophyte, the nutritive tissue that was part of the female gametophyte, and the embryo that represents the new sporophyte generation.

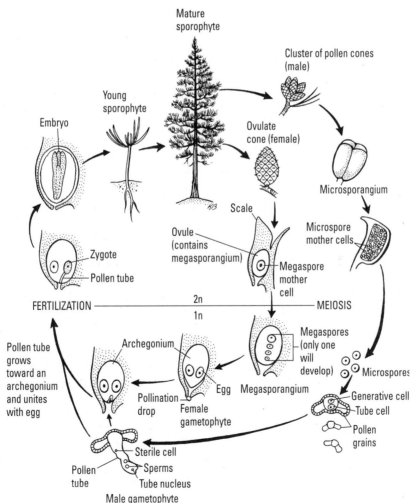

Figure 16-4: Life cycle of pine.

Gnetophytes: Gnetophyta

Of all the gymnosperms, botanists believe that gnetophytes are most closely related to the flowering plants. The relationship between gnetophytes and angiosperms is supported by comparisons of DNA sequences and by some structural characteristics, such as the presence of xylem vessels.

Like the cycads and ginkgos, gnetophytes were much more diverse in the distant past — today just three groups of gnetophytes remain:

- The genus *Gnetum* is the most common member of a group of broad-leaved tropical plants, many of which are vines. The leaves of these plants have branching veins very similar to those of the flowering plants. *Gnetum* produces pollen and seeds in bisexual cones.

- *Ephedra,* also called Mormon tea or by its Chinese name *ma huang,* are brushy shrubs that grow in desert areas. The plants produce two or three tiny leaves at each node along the stem, but the leaves turn brown quickly and fall off, giving the plant a jointed stick-like appearance that explains why it's also sometimes called a *joint fir.* The stems of these plants are green and photosynthetic. *Ephedra* is well-known in medical circles because it produces *ephedrine,* which is a powerful decongestant.

- *Welwitschia* is definitely the strangest of the gnetophytes. Each plant produces two large, strap-like leaves that grow continuously throughout the life of the plant. The leaf meristems are located at the base of each leaf, near the large cup that forms in the center of the plant. So, although the leaves may get split and tattered at their tips, they continue to grow from the base. *Welwitchia* is dioecious, so some plants produce female cones while others produce male cones. The cones are produced around the rim of the central cup.

Chapter 17

Say It with Flowers: Angiosperms

In This Chapter
▶ Examining the evolution of flowering plants
▶ Digging into flowering plant diversity
▶ Checking out pollination strategies

Flowering plants are the most abundant group of plants on planet earth, providing food, clothing, and building materials to people, as well as being the source of homes and food for many other species. And beyond what flowering plants do for people, they're fascinating and beautiful to look at. This chapter discusses the main characteristics of flowering plants, explores the mystery of their origins, and gives you an introduction to their incredible diversity.

Delving into Flowering Plants: Anthophyta

Flowering plants are intimately woven into your daily life. Even if you haven't stopped to smell the roses today or given any as a gift, you've almost certainly eaten food, worn clothing, and used objects that came from flowering plants.

Flowering plants, called *angiosperms,* are the most abundant and diverse group of plants on planet earth. Figure 17-1 gives you just a peek into the diversity of plants found in this group.

Searching for the origin of the angiosperms

The angiosperms came blazing into the fossil record of the Cretaceous period (144 to 65 million years ago), with so many diverse groups suddenly appearing that the evolutionary origin of the angiosperms has been a puzzle to botanists ever since Charles Darwin called it an "abominable mystery."

The origin of angiosperms puzzles botanists for two reasons:

✓ Botanists think the ancestors of angiosperms and gymnosperms diverged from each other back in the Paleozoic (about 305 million years ago).

✓ Fossils that look like modern angiosperms don't appear until the Cretaceous Period — leaving a gap of over 100 million years.

Figure 17-1: Angiosperm diversity.

a. Water lily
b. Pepper
c. Tulip tree
d. Buttercup
e. Sunflower
f. Wheat

Chapter 17: Say It with Flowers: Angiosperms **277**

The changes that occurred in the angiosperm lineage during all that time are largely still a mystery to botanists. Some botanists propose early angiosperm evolution must have occurred in places where fossils weren't well preserved. Others think the fossil record is fairly accurate, and angiosperms just became wildly successful very suddenly. Botanists continue to try to answer this question both by studying living angiosperms and by continuing to study the fossil record.

Comparing DNA for clues

Fossils aren't the only way to explore the past anymore. Now that scientists can read the DNA of organisms, they can compare the genes of living organisms and use the information to build family trees, called phylogenies, of life on earth. By grouping together organisms that are the most similar, scientists essentially work backwards down along the trunk of the tree to figure out which living organisms are most closely related to the ancestors at the base of the tree.

Based on a recent comparison of the DNA of living angiosperms, botanists have created the following hypothesis for the evolution of angiosperms:

- Modern angiosperms that are descendants of the earliest angiosperms form several branches at the base of the angiosperm tree, and so are called *basal angiosperms*. These plants, which botanists think may be similar to the earliest angiosperms, include a shrub called *Amborella*, water lilies (*Nymphaea*, see Figure 17-1A), and star anise (*Illicium*).

- After the evolution of the basal angiosperms, three more groups of angiosperms appeared, although botanists are still studying the relationships between these three groups:

    - The *magnoliids* is a group of plants that includes magnolias (*Magnolia*, Figure 17-2), tulip trees (*Liriodendron*, Figure 17-1C), and black pepper (*Piper,* Figure 17-1B).

    - The eudicots, which include buttercups (Figure 17-1D), sunflowers (Figure 17-1E), and maple trees.

    - The monocots, which include grasses (Figure 17-1E), lilies, and palms. (For a comparison of dicot and monocot characteristics, head to Chapters 4 and 5.)

For a long time, botanists thought that the flowers of *Magnolia*, such as that in Figure 17-2, represented the earliest angiosperm flowers. Recent evidence (see sidebar "The abominable mystery" for details) contradicts this view.

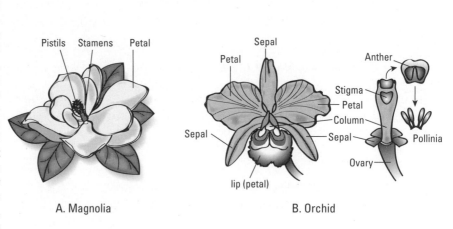

Figure 17-2: Primitive versus advanced characteristics in angiosperms. Magnolia flowers have characteristics, such as unfused parts, that are seen in early angiosperm flowers, whereas orchid flowers have many fused parts.

Discovering the unique characteristics of angiosperms

Despite the mystery surrounding their origins, one thing is clear: Angiosperms became incredibly successful, perhaps because they combined the advantages of seeds with the evolutionary innovations of flowers and fruits. Angiosperms and gymnosperms are closely related, and they have many characteristics in common (see Chapter 16).

However, angiosperms also have many unique traits that unite them into the phylum Anthophyta (also called Magnoliophyta and Angiospermae):

- ✓ **Ovules enclosed in carpels:** Over time, carpels became modified to have the three parts now seen in modern flowers: the ovary, style, and stigma (see Chapter 5).

- ✓ **Pollen germinates on the stigma, rather than directly on the ovule as it does in gymnosperms:** In order for fertilization to happen, the pollen tube must grow down through the style and deliver sperm to the ovules inside the ovary. By putting some distance between the pollen and the ovules, angiosperms gained more control over fertilization. For example, in some species, if pollen from the same plant lands on the stigma it won't germinate at all, which prevents self-pollination. By making sure that fertilization comes from a separate individual, genetic diversity in the offspring is increased.

✔ **Smaller gametophytes:** Just as in gymnosperms, the gametophyte generation of angiosperms is very small and completely enclosed and protected by sporophyte tissue. But in angiosperms, gametophytes are even smaller.

- Female gametophytes, called megagametophytes or the *embryo sac,* are made of just seven cells and eight nuclei. (One cell has two nuclei.)
- Male gametophytes, called microgametophytes or pollen, are made of just three cells: one tube cell and two sperm cells. The male gametophyte, with a germinating pollen tube, is shown in Figure 17-3.

✔ **Development of endosperm to support the embryo:** Angiosperms have a unique type of fertilization called *double fertilization.*

- One sperm cell fertilizes the egg to produce the embryo.
- One sperm cell fertilizes two nuclei called *polar nuclei,* to produce the *endosperm.* Endosperm contains three sets of chromosomes, one from each polar nucleus and one from the sperm cell, so it is *triploid.*

Looking at the life cycle of angiosperms

As in gymnosperms, the sporophyte is the dominant generation in angiosperms. Figure 17-4 illustrates the life cycle of angiosperms in a plant that has both male and female parts in the flower:

1. **The male gametophyte develops inside the anther of the flower.**

   Microspore mother cells (microsporocytes) inside the anther divide by meiosis to produce haploid microspores. Then the microspores divide by mitosis to produce the male gametophyte inside the pollen grain. Initially, the male gametophyte consists of two cells: the *tube cell* and the *generative cell.* The tube cell produces the pollen tube. The generative cell will divide again to produce two sperm cells.

2. **The female gametophyte develops inside the ovule of the flower.**

   The megaspore mother cells (megasporocytes) inside the ovule divide by meiosis to produce four haploid megaspores. Three megaspores break down, and then the surviving megaspore divides by mitosis to produce the female gametophyte, which is also called the embryo sac.

3. **Pollination occurs when pollen lands on the stigma.**

   The pollen tube grows down through the style and then through the micropyle of the ovule, delivering sperm cells to the female gametophyte.

4. **Double fertilization occurs.**

   One sperm cell fertilizes the egg, forming the diploid zygote. The other sperm cell combines with the two polar nuclei, forming the triploid endosperm.

5. **The ovule develops into a seed, and the ovary becomes a fruit.**

   The zygote divides by mitosis into a multicellular embryo that has one or two cotyledons. In dicots, the cotyledons act as food reserves for the embryo; in monocots, the endosperm acts as the food reserve.

6. **When environmental conditions are right, the seed germinates and begins to grow.**

   The embryo develops into a new sporophyte.

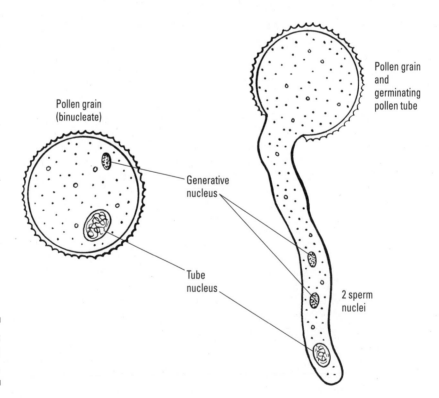

Figure 17-3: Growth of a pollen tube.

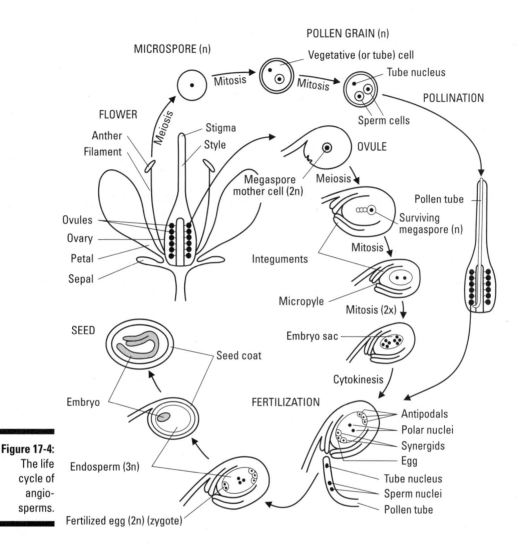

Figure 17-4: The life cycle of angiosperms.

Exploring Angiosperm Diversity

More than 250,000 species of angiosperms populate the landscape today. Although all angiosperms make flowers, the appearance of the flowers range from flowers completely lacking petals, such as those of grasses (Figure 17-1F), to showy flowers with lots of parts, such as those of water lilies (Figure 17-1A).

Basal angiosperms

The oldest groups of angiosperms represent a very small percentage of the total species that exist today, but they're important because they provide clues to what the earliest angiosperms looked like.

Early angiosperm flowers, such as those of the basal angiosperms and magnoliids, have radial symmetry. When something is radially symmetrical, you can imagine cutting it into slices like you'd cut a pie, and the slices all look the same. For example, bicycles wheels are radially symmetrical. More recently evolved angiosperms, such as eudicots and monocots, often have flowers with bilateral symmetry. When something is bilaterally symmetrical, its two halves mirror each other. So, if you imagine cutting it down the middle, the two halves look the same. For example, your body is bilaterally symmetrical. Botanists call flowers that are radially symmetrical *actinomorphic,* or regular, and flowers that are bilaterally symmetrical *zygomorphic,* or irregular.

Flowers of the basal angiosperms are typically bisexual with lots of separate parts that are spirally arranged. The flowers show radial symmetry and are typically insect-pollinated. Their pollen grains have one opening through which the pollen tube can grow. Some basal angiosperms don't make vessels in their xylem.

Magnoliids

Magnoliids make up about 2 percent of angiosperms. Some magnoliids that may be familiar to you are magnolias (see Figure 17-2) and tulip trees (see Figure 17-1C).

The flowers of magnoliids have several characteristics in common with those of the basal angiosperms, such as making numerous, separate parts that are spirally arranged. Some magnoliids also make leaf-like stamens and unsealed carpels. Their pollen grains have one opening.

True dicots

True dicots are the largest group of angiosperms, containing about 75 percent of all angiosperm species. True dicots include familiar garden plants beans, cucumbers, and broccoli; wildflowers, such as poppies, asters, and geraniums; important ornamental and food-producing plants, such as apples and cherries; and trees, such as oak and maple.

Botanists call this group the *eudicots*, which means true dicots, to separate them from the magnoliids. The magnoliids also have two cotyledons, but evolved separately from the eudicots. Although these two groups are separate in terms of evolution, many classical taxonomic schemes place them together in one class (Magnoliopsida or Dicotyledones).

There are so many different kinds of eudicots, in fact, that it would take a whole book just to describe them all. As a way of at least introducing you to this group, I present a few examples of various eudicot flowers. Also, don't forget that for the general vegetative and reproductive characteristics of dicots, you can refer to Chapters 4 and 5.

Figure 17-1D shows a flower from the buttercup family. This family, the Ranunculaceae, belongs to one of the earliest eudicot groups to evolve. The flowers have many separate male and female parts, just like those of the basal angiosperms and the magnoliids. One difference, however, is that the pollen grains of buttercups have three openings through which the pollen tube can emerge, rather than just one opening.

True dicot pollen has three openings for the pollen to grow out, which is one of the characteristics that separates this group from the other groups of angiosperms.

Figure 17-5 shows a flower from the legume family, or Fabaceae, which botanists used to call the Leguminosae. The legume family is the third largest flowering plant family and has huge importance to human agriculture. Peas, beans, and peanuts all belong to this family.

Some familiar legumes, such as the one shown in Figure 17-5, make a special type of flower that's sometimes called a *flag flower* (or papilinioid flower):

- Flag flowers are bisexual flowers with five fused sepals and five petals that have unique shapes:
    - One petal, called the banner or standard sticks up above the flower.
    - Two petals, called the wings, stick out to the sides.
    - Two petals are fused together to form a cupped structure, called the keel, at the bottom of the flower.
- Most legume flowers have ten stamens, nine of which are fused together and one of which is separate.

Figure 17-6 shows a flower from the spurge family, or Euphorbiaceae. This family is one of the largest plant families and is known for plants that produce milky sap. Many spurges are succulent, resembling cacti.

Spurge flowers lack petals and many lack sepals as well, so that flowers may consist of just male or female parts such as the one in Figure 17-6. The flowers are unisexual.

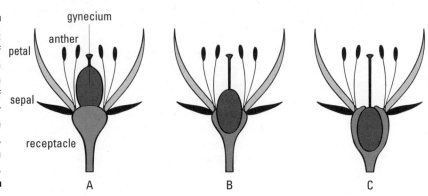

Figure 17-5: Parts of a legume flower. The five petals of this flower have unique shapes, forming a flag flower.

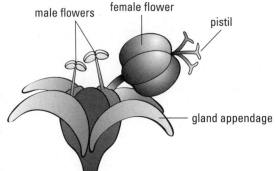

Figure 17-6: Parts of a spurge flower.

Tomatoes, potatoes, chile peppers, tobacco, jimson weed, and deadly nightshade all belong to the nightshade family, or Solanaceae. The family is famous for producing chemicals called alkaloids, many of which are extremely toxic. Flowers in this family often have fused petals that form a tube with a pleated appearance. The flowers also have stamens that are partially fused to the petals. Botanists refer to stamens like these as *epipetalous*, because *epi-* means upon, and the stamens are upon the petals.

The sunflower family, or Asteraceae, is the largest dicot family and includes many beautiful wildflowers such as sunflowers, and the familiar pest of lawns and gardens, the dandelion.

The flowers of the Asteraceae form a characteristic inflorescence called a head that looks like a single flower but is actually a composite of many flowers. The part of the head that looks like petals is made up of *ray flowers*. The center of the head is made up of *disk flowers*. Each disk flower has a tubular corolla. The stamens in this family form a tube around the style, with the anthers fused together. The heads of asters may have both ray and disk flowers, like in a sunflower, just disk flowers like in thistles, or just ray flowers, like in dandelions.

Monocots

Monocots make up about 23 percent of angiosperm species. Monocots include orchids, grasses, lilies, and palm trees.

For the general vegetative and reproductive characteristics of monocots, see Chapters 4 and 5. As examples of the diversity of this group, this section presents a few monocots that have very different looking flowers.

Members of the lily family, or Liliaceae, such as the flower in Figure 17-7, have classic monocot flowers with parts in threes. Members of this family typically have bisexual flowers with parts in multiples of three, such as six tepals (or three sepals and three petals), six stamens, and three fused carpels. The flowers show radial symmetry and produce pollen grains with one opening. Most members of this family are herbs.

Although monocot pollen has one opening, the structure of monocot pollen is different from the pollen of the basal angiosperms and magnoliids. Based on all their characteristics, including DNA comparisons, botanists think that monocots are a unique group of angiosperms that evolved later than the basal angiosperm group and that is most closely related to the eudicots and magnoliids.

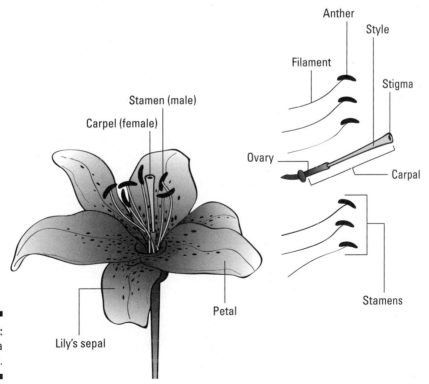

Figure 17-7: Parts of a lily flower.

Orchids are some of the most exotic and beautiful flowers in the world. They're prized as ornamentals and also include some important agricultural plants, like vanilla. Members of this family, such as the flower in Figure 17-2, often have very unusual looking flowers, some of which mimic insects or bats. (For more details, see the next section on pollination ecology).

Orchid flowers are typically bisexual and bilaterally symmetrical, having three sepals and three petals, with one petal modified to form a landing platform for pollinators called a *labellum*. The flowers have one or two stamens and three fused carpels. The style, stamens, and stigma are also fused together to form a central column. The pollen grains have one opening are packaged into bags called *pollinia*. Orchid seeds are tiny and lack endosperm, and they can't even germinate unless they establish a symbiotic relationship with a mycorrhizal fungus.

Many orchids are epiphytic, growing on other plants. These orchids have swollen stems called *pseudobulbs* that help store water. They also have aerial roots covered with a layered epidermis called *velamen* that helps orchids collect water and minerals from the air.

Grass flowers are usually bisexual, with zero to three tepals. Tepals are sometimes modified to scale-like structures called *lodicules* at the base of the flower. The flowers have one, two, three or six stamens, and two to three carpels fused to make one pistil. The stigma is feathery, which is useful for catching airborne pollen.

The flowers are arranged in inflorescences called *spikelets*. Each flower sits in a pair of bracts, forming a *floret*. Each floret has a lower bract called a *lemma* and an upper bract called a *palea*. One or more florets grow above two bracts called *glumes,* forming the spikelet. The main axis of the spikelet that supports the florets is the *rachilla*.

Grass leaves occur in pairs in a two-ranked arrangement: The leaves alternate and have bases that wrap around the stem so that one leaf sits slightly above the other, emerging from its open sheath. A small flap of tissue called a ligule is attached to the leaf base where it wraps around the stem.

Thinking about Pollination Ecology

One really cool thing about angiosperms is their interactions with animals. Angiosperms control animal behaviors, getting animals to bring them pollen by attracting them with certain colors and scents and offering them rewards of food. Angiosperms also get animals to disperse their fruits, seeds, and pollen by hitching a ride on their surfaces or offering them food. But wait a minute! Perhaps it's the other way around. Maybe the animals caused the angiosperms to serve the needs of the animals by dispersing pollen and seeds. Actually, scientists think it's a bit of both — that over time, the interactions between angiosperms and animals have affected the evolution of both groups.

Coevolution is when two species affect the evolution of each other, so that both groups develop traits in response to the other. (For more details on evolution and how it works, see Chapter 13.) The coevolution of angiosperms and animals is part of the reason that angiosperms are so colorful and diverse. Flowers have changed over time to be attractive to certain animals, and not others. And animals have changed to efficiently gather resources, such as nectar from flowers.

You can make a pretty good guess about which plants are getting animals to work for them, by checking the plants for animal signals and rewards. Plants that lure animal pollinators typically have bright colors and intriguing scents and may offer a food reward such as nectar. Wind-pollinated plants, by contrast, may seem drab — no showy flowers, just reproductive parts hanging in the breeze, but if you have allergies, you may definitely notice the wind-pollinated plants because they tend to put out lots of pollen to make sure that they get the job done!

Plants that lure animals to disperse their fruits and seeds typically fall into one of two camps: those that offer juicy rewards like fleshy fruits, and those that sneakily hitch a ride on fur, hair, or feathers, using sticky barbs and spines. Plants that rely upon the wind to do the work make light, dry seeds and fruits that can float easily on the breeze. Many of them make wings or parachutes to help carry their reproductive structures on the breeze.

By attracting animal pollinators, plants help make sure that they receive pollen from diverse mates, helping to increase genetic diversity. When an animal visits a flower, pollen is transferred to its body. Then, when the animals move to a different individual, some of the pollen from the previous flower rubs off, achieving *cross-pollination*.

Flowers and their pollinators have evolved very fine-tuned relationships:

- **Flowers that are pollinated by insects provide a food reward and have colors, scents, and shapes that attract certain insects:**
  - *Bee-pollinated flowers are usually bright blue or yellow because those are colors that bees see well, and they have a sweet fragrance to attract the bees.* Bees can also see something you can't — ultraviolet light — so bee-pollinated flowers look different to the bees than they do to you. If you were to shine a UV light on bee flowers, invisible patterns would be revealed to you, patterns that guide the bees to the center of the flower to find their rewards of nectar and pollen. Bees don't see red, so red flowers look black to them and don't attract the bees' attention.

- *Fly-pollinated flowers tend to be dark red or brown colors and have smells like rotten meat or animal waste.* (I personally have smelled fly-pollinated flowers that reminded me of urine.) Because of their rotten meat appearance and smell, some fly-pollinated flowers are called *carrion flowers*. Flies are attracted to the flowers to lay their eggs and in the process, pick up pollen for dispersal.

- *Beetle-pollinated flowers are usually white or dull colors and have strong fruity or spicy odors.* The flowers provide the beetles with food, either as pollen or by making special food-storage cells on the petals.

- *Some wasp-pollinated orchids look like certain female wasps and trick the male wasps into attempting to mate with them!* For more on this strange occurrence, see Chapter 21.

- *Butterfly-pollinated flowers are usually brightly-colored, tube-shaped flowers that have a sweet smell.* Butterflies have long, rolled-up tongues, so they can probe deeply into tube-shaped flowers for their nectar reward. Unlike bees, butterflies can see the color red so many butterfly-pollinated flowers are red, orange, or yellow.

- *Moth-pollinated flowers are similar in shape and smell to butterfly-pollinated flowers, but are usually white.* That's because moth-pollinated flowers are typically open at night when moths are more active and white is a good color for being seen at night.

- **Bird-pollinated flowers are usually bright red or yellow but don't have much scent.** That's because birds have great vision, but a poor sense of smell. Bird-pollinated flowers may be tubular like butterfly flowers because bird beaks can probe deeply into flowers. Birds, especially hummingbirds, use up a lot of energy while visiting flowers, so the flowers provide lots of nectar to make sure the birds keep coming back for more.

- **Bat-pollinated flowers tend to be single large flowers or clusters of many flowers that are white or light-colored and have a sweet fruity smell or a musty, bat-like smell.** Like moth flowers, bat-pollinated flowers open and release their scents at night. The light colors help make the flowers visible at night. Bats shove their whole heads into the flowers in order to reach the nectar with their tongues.

Part V
Plants and People

The 5th Wave — By Rich Tennant

"Good news, we found a plant that cures baldness."

In this part . . .

The lives of plants and people are interconnected in many ways. Plants play many essential roles in natural ecosystems, from making food to recycling carbon to providing homes for other species. Plants also provide people with useful products, such as food, oxygen, medicines, clothing, and wood.

In this part, I explore the many ways in which plants support the lives of people, from their role in nature to the useful products they provide.

Chapter 18
Making Connections with Plant Ecology

. . . .

In This Chapter
▶ Mastering the fundamentals of ecology
▶ Investigating plant interactions
▶ Exploring plant communities

. . . .

*P*lants are an incredibly important part of life on earth. Plants make the food upon which all organisms rely. They provide homes to many other species, and they are part of the structure of ecosystems. This chapter covers the fundamental principles that underlie the interactions between all species and explores the importance of plant communities in the environment.

Exploring Ecosystems

All life on earth is interconnected — organisms depend on each other for food and are impacted by each other's actions. Life depends on the environment too, for water, light, minerals, and a place to live.

An *ecosystem* is a group of living and nonliving things interacting with each other in a particular environment. Ecosystems can be large or small, from the entire web of life interacting with each other and the surface of the earth, called the *biosphere*, to a tiny ecosystem in a tide pool along the seashore.

Whatever their size, every ecosystem includes certain components:

 ✓ **Abiotic factors are the nonliving parts of the ecosystem.** Abiotic factors include things like sunlight, water, soil, air, and dead materials.
 ✓ **Biotic factors are the living components of the ecosystem.** Scientists organize the biotic factors into two categories (shown in Figure 18-1):
   • *Populations* are groups of the same species living in the same place. A forest, for example, may include a population of oak trees,

a population of salamanders, and a population of squirrels. Even very similar organisms are considered different populations if they aren't the same species.

- *Communities* are groups of one to many species that live in the same place. All the populations of different species that live in the same forest make up the forest community.

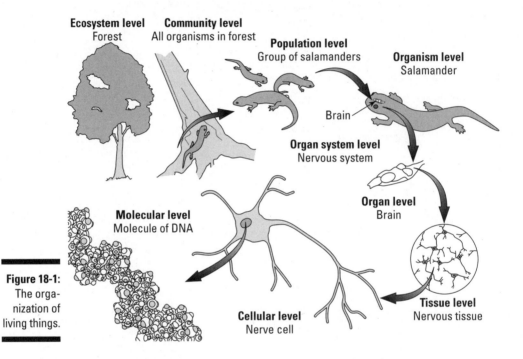

Figure 18-1: The organization of living things.

Scientists who study ecosystems, called *ecologists,* are interested not only in the components of the ecosystem, but also in how all the components interact with each other. *Ecology* is the study of how organisms interact with each other and their environment.

Figuring out your job description

Every species in an ecosystem plays a certain role in that system — it lives in a certain area, uses resources, and either eats or is eaten by other species.

The particular role a species plays in an ecosystem — including where it lives and which resources it uses — is called its *ecological niche*. The particular place where a species lives is its *habitat*.

Organisms have many different ways of getting it, but they all need food. Even plants need food — they just make it themselves. All organisms need food because it provides the energy and matter necessary for growth, repair, and reproduction. For more on how individual organisms use food, head to Chapter 6.

How an organism gets its food is a big part of its niche. Ecologists commonly divide the organisms in an ecosystem into three categories, called *trophic levels,* based on how they get their food:

- ✓ **Producers make their own food.** Plants, algae, and photosynthetic bacteria are all producers that use energy from the sun to combine carbon dioxide and water, forming carbohydrates by the process of photosynthesis. Producers can also be called *autotrophs.*

- ✓ **Consumers eat other organisms.** Consumers can be further divided up based on what type of organism they eat.

    • *Primary consumers* eat producers. Because producers are mainly plants, primary consumers are also called *herbivores* (plant-eating animals).

    • *Secondary consumers* eat primary consumers. Because primary consumers are animals, secondary consumers are also called *carnivores* (meat-eating animals).

    • *Tertiary consumers* eat secondary consumers, so they are also carnivores.

- ✓ **Decomposers break down dead matter.** True decomposers, such as bacteria and fungi, release enzymes onto dead matter, breaking it down into solution, and then absorbing it as food. Other organisms, such as scavengers and detritivores, also eat dead matter, but they ingest larger pieces and break them down internally. Scavengers, such as crabs and vultures, often feed on the remains of carcasses left by consumers, while detritivores, such as worms, ingest small pieces of *detritus,* the collection of random dead matter in an ecosystem.

Going with the flow of energy

Plants and other photosynthetic organisms are fundamental to most ecosystems on earth because they can capture energy and matter and store it in food molecules. All other organisms in an ecosystem ultimately rely upon the food made by plants and other autotrophs. Primary consumers eat plants, then secondary consumers eat primary consumers, and so on, creating *food chains,* such as the one in Figure 18-2, that show how each organism gets its food.

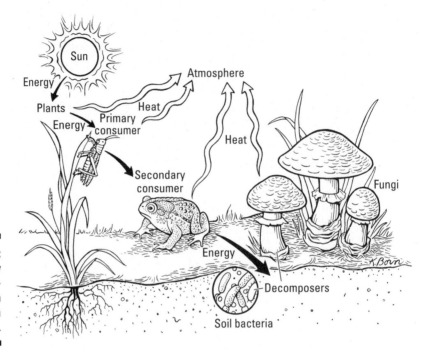

Figure 18-2: Energy flow in ecosystems shown through a food chain.

In an ecosystem, food chains branch and intersect with each other to create a complex web of interactions called a *food web*. Food chains and food webs show the flow of energy and matter through the different trophic levels of an ecosystem.

Living by the laws of energy

The sun is the ultimate source of energy for almost all life on earth — without the energy from the sun, life on earth just couldn't exist. Energy from the sun is captured by plants and other producers that make the food for all living things. Even the types of energy that humans rely upon, such as electricity and gasoline, can ultimately be traced back to the sun. For example, most of the oil and coal people rely upon today were made from the remains of plants that grew some 350 million years ago during the Carboniferous period. The sun radiates a constant flow of energy upon the earth, and plants are first in line to capture it and make it usable for life. Understanding the movement of energy from the sun through ecosystems will be much easier if you know a few rules about the nature of energy.

Four really important energy principles form the foundation of interactions in ecosystems:

- **Energy can't be created or destroyed.** This statement represents a fundamental law of the universe, called the *first law of thermodynamics*. The consequence of this law is that every living thing has to get its energy from somewhere; they can't make it for themselves. Even plants, which make their own food, can't make their own energy — they capture energy from the sun and store it in the food they make.

- **When energy is moved from one place to another, the energy is *transferred*.** When a primary consumer eats a producer, the energy that was stored in the body of the producer is transferred to the primary consumer. When describing energy transfers, be sure to say where they energy is coming from and where it's going to.

- **When energy is changed from one form to another, the energy is *transformed*.** When plants do photosynthesis, they absorb light energy from the sun and convert it into the chemical energy stored in carbohydrates. So, during photosynthesis, light energy is transformed into chemical energy. When describing energy transformations, be sure to state the form of energy before and after the transformation.

- **When energy is transferred in living systems, some of the energy is transformed into heat energy.** This statement is one way of representing another law of the universe, called the *second law of thermodynamics*. The impact of this law on ecosystems is that no energy transfer is 100 percent efficient. Once energy is transformed to heat, it's no longer useful as a source of energy to living things.

Stacking up the costs of energy transfer

Energy from the sun transfers through the trophic levels of food webs, supplying living things with the energy they need. However, the amount of energy that's available to life decreases as it moves through the trophic levels. Every organism constantly transfers energy for its own needs, and every energy transfer comes with a cost of energy that's transferred to the environment as heat. As an example, think about your own body as you exercise — you heat up and begin to sweat. Your body heats up from the all the energy transfers your cells are doing in order to provide energy to your muscles. The heat energy then transfers to your skin and out to the air around you. This example is pretty dramatic — you can actually feel the heat radiating off a person who has been exercising — but the same thing is going on all the time, in your body and in the bodies of living things all around you. Even plants, which can't get up and run around, are constantly doing energy transfers inside their cells, causing some energy to become heat that's then released to the environment.

So, if the organisms in each trophic level use some of the energy they capture for their own functions, causing some of the energy to be released as heat to the environment, then the total amount of energy available as you move through the trophic levels becomes less. This decline in the amount of available energy can be represented by an *energy pyramid*, also called a *trophic pyramid*. Energy pyramids like the one in Figure 18-3 show how the available energy decreases as it moves through the trophic levels in an ecosystem:

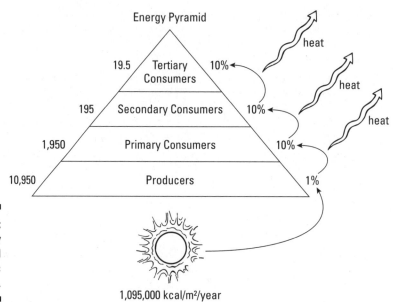

Figure 18-3: The energy pyramid (trophic pyramid).

- ✔ **Producers capture about 1 percent of the energy available from the sun and store it in the molecules that make up their bodies.** Producers grow, transferring much of their stored energy into ATP for cellular processes (see Chapter 6 for more on ATP) and the molecules that make up their bodies. As producers transfer energy for growth, some energy is also transferred to heat that's released to the environment. The sun releases about 1,095,000 kilocalories (kcal or C) of energy per meter squared per year. Of that total energy, only about 10,950 kcal is stored in the bodies of producers.

- ✔ **Primary consumers get about 10 percent of the energy that was captured by the producers.** Primary consumers eat the producers, using the molecules from their bodies as food. Just like producers, the primary consumers grow, transferring energy from food into ATP for cellular processes and into the molecules that make up their bodies. As primary consumers transfer energy for growth, some energy is also transformed to heat that's released to the environment. Of the total energy released by the sun, only about 1,950 kcal is stored in the bodies of the primary consumers.

- ✔ **Secondary consumers get about 10 percent of the energy that was captured by the primary consumers.** Secondary consumers eat primary consumers, using the molecules from their bodies as food. Secondary consumers grow, transferring energy from food into ATP for cellular processes and into the molecules that make up their bodies. As secondary consumers transfer energy for growth, some energy is also transformed to heat that's released to the environment. Of the total energy released by the sun, only about 195 kcal is stored in the bodies of the secondary **consumers.**

Chapter 18: Making Connections with Plant Ecology

- ✔ **Tertiary consumers get about 10 percent of the energy that was captured by the secondary consumers.** Tertiary consumers eat secondary consumers, using the molecules from their bodies as food. Tertiary consumers grow, transferring energy from food into ATP for cellular processes and into the molecules that make up their bodies. As tertiary consumers transfer energy for growth, some energy is also transformed to heat that's released to the environment. Of the total energy released by the sun, only about 19.5 kcal is stored in the bodies of the secondary consumers.

- ✔ **Decomposers use the dead matter from every level of the pyramid as their source of food.** As organisms die and some of their remains become part of the environment, decomposers use these remains as their source of energy and matter for growth. These organisms also transfer energy from food into ATP for cellular work and into the molecules that make up their bodies. As they transfer energy for growth, some energy is transformed to heat that's released to the environment.

Ecologists say that energy flows through ecosystems. Energy enters ecosystems from the sun, passes through the bodies of living things, and then flows back out the environment as heat released to the atmosphere. Ultimately, all the heat released by the bodies of organisms is transferred from the atmosphere back out to outer space.

You may have heard of organisms that are *top predators* — in other words, the organism at the top of a food chain. Top predators usually do not go beyond the level of tertiary consumers because the amount of energy available to support life becomes less as you move up the energy pyramid. Past the tertiary consumer level, too much energy has been depleted from the system to support higher trophic levels.

The amount of energy at each trophic level in proportion to the next trophic level is called *ecological efficiency*. The general rule that ecologists use for ecological efficiency is the *10 percent rule*.

You may hear people, even scientists, say that "energy is lost" or "energy is lost as heat." These statements can be confusing because they make it sound like energy disappears somehow. But, you know from the first law of thermodynamics that energy can't be destroyed or disappear. The correct interpretation of statements like these is that useful energy is lost from the system as it is transformed into heat. In other words, once energy is transformed to heat, organisms in an ecosystem can't use it as a source of energy for growth. So, a scientist, even your teacher, may say something about energy being lost. They probably know what they mean, but it may get confusing for you. So, take my advice and never use the words lost, disappear, destroyed, or created when you're talking about energy. Use the words transfer and transformed instead, and you'll avoid a great deal of confusion.

Going round and round with matter cycles

Almost all the matter on planet earth today, including the molecules that make up your body and the bodies of the living things all around you, has been here since the earth first formed. Think about it — energy constantly comes in from the sun, but the earth isn't constantly being bombarded with asteroids. So, where do living things get the raw materials to build their bodies? By recycling! All the carbon, hydrogen, oxygen, nitrogen, and other elements that make up the molecules of living things have been recycled over and over throughout time. When decomposers break down the bodies of dead organisms, they release atoms back to the environment in forms that other living things can use.

Ecologists say that matter cycles through ecosystems. Scientists track the recycling of atoms through cycles called *biogeochemical cycles* (*bio-* because the recycling involves living things, *geo-* because it involves the earth, and *chemical* because it involves chemical processes).

The carbon cycle

Life on earth is made of carbon — the proteins, carbohydrates, and fats that make up the bodies of living things all have a carbon backbone. Carbon becomes available to ecosystems when producers, such as plants, capture it from the atmosphere, using it to build carbohydrates in the process of photosynthesis. (For the details on photosynthesis, check out Chapter 7.) Producers capture carbon and incorporate it into food molecules in a process called *carbon fixation*. Once producers produce food molecules, organisms recycle the carbon through ecosystems, moving it through food webs and back to the atmosphere, forming a pathway of carbon called the *carbon cycle*. Figure 18-4 presents the key processes that transfer carbon through the carbon cycle:

- **Producers fix carbon from the environment and incorporate it into food molecules.** Most ecosystems rely upon producers such as plants to fix carbon using photosynthesis. During the process of photosynthesis, plants make glucose ($C_6H_{12}O_6$) out of carbon dioxide (CO_2). Plants then use the food they make as a source of matter to build the molecules that make up their own bodies.

- **Consumers and decomposers eat other organisms, transferring carbon into their own bodies.** Consumers and decomposers mostly take in proteins, carbohydrates, and fats from their food. Then, they rearrange the molecules from their food into the molecules they need for their own bodies.

- **When all living organisms break down food molecules for energy, they release carbon back to the environment as carbon dioxide.** Many organisms use cellular respiration to break down food molecules and transfer the energy from food to ATP for use in cellular processes. When food molecules are broken down to allow energy transfer, cells rearrange the bonds in the molecules, forming carbon dioxide (CO_2) and water (H_2O) as waste. These molecules have very little available energy, so cells release them back to the atmosphere as waste.

Although most people know that animals release carbon dioxide to the atmosphere, they forget that other kinds of organisms — including plants! — also release carbon dioxide molecules from the breakdown of food. It's hard to think of plants as releasing carbon dioxide because they take up so much of it as part of photosynthesis. But the truth is plants make food and use it too! For more on cellular respiration in plants, check out Chapter 8.

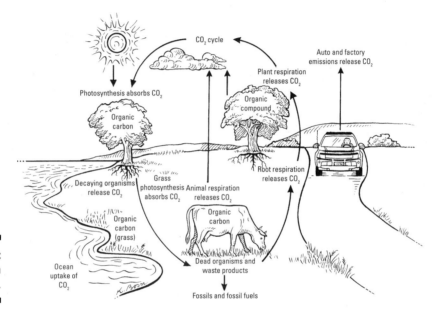

Figure 18-4: The carbon cycle.

As carbon cycles through ecosystems, the balance of carbon can shift between locations.

Carbon sinks store carbon, while *carbon sources* release carbon to the environment. Carbon sinks can store carbon in the environment for long or short periods of time:

- The bodies of living things store carbon for relatively short periods of times.
- Fossil fuels contain carbon that was stored in the bodies of living things long ago and then trapped in a way that the proteins, carbohydrates, and fats were converted to coal, oil, and natural gas deposits.
- The world's oceans store large amounts of carbon dioxide dissolved as gas in the water.
- Mineral forms of carbon, such as calcium carbonate, store carbon in the shells of some living things. Fossil shells can also store significant amounts of carbon, like those that make up the White Cliffs of Dover in England.

Human activities, particularly those since the Industrial Revolution, are having a big impact on the global carbon balance

- **When people cut down forests, a process called deforestation, they eliminate carbon sinks.** In addition, if the wood is burned, the burning becomes a source of carbon to the atmosphere.
- **The burning of fossil fuels is also a source of carbon to the atmosphere.** As people burn fossil fuels, this stored carbon is being rapidly released back to the atmosphere as carbon dioxide, causing the concentration of carbon dioxide in the atmosphere to increase to its highest levels in recorded history.
- **Increases in global temperature are limiting the ability of the ocean to act as a carbon sink.** Warm water holds less carbon dioxide than cold water, so some of the carbon stored in the ocean may be released back to the atmosphere as ocean temperatures rise as a result of global warming.

The nitrogen cycle

Nitrogen is essential to life because cells need it to build the amino acids that make up proteins and to make DNA and RNA. (For more on the structure of these molecules, see Chapter 2)

Biological nitrogen fixation occurs when nitrogen-fixing bacteria convert nitrogen gas from the atmosphere into forms of nitrogen that can be used by living things. Once nitrogen fixation occurs, organisms recycle the nitrogen through ecosystems, moving it through food webs and back to the atmosphere, forming a pathway of nitrogen called the *nitrogen cycle*.

Figure 18-5 presents the key processes that transfer nitrogen through the nitrogen cycle:

- **Nitrogen fixation changes atmospheric nitrogen into forms of nitrogen that are usable by living things.** Nitrogen gas in the atmosphere can't be incorporated into the molecules of living things, so ecosystems depend upon the activity of bacteria that live in the soil and in the roots of plants. These *nitrogen-fixing bacteria* convert nitrogen gas (N_2) into forms like ammonium or nitrate ions (NH_4^+ or NO_3^-) that can be used by living organisms. Some nitrogen fixation also occurs by lightning strikes and by processes in factories that produce chemical fertilizers for plants. However, the nitrogen fixation that occurs from lightning strikes isn't enough to supply ecosystems with all the nitrogen they need, and industrial nitrogen fixation requires a lot of energy.
- **Producers get nitrogen by absorbing ammonia and nitrate along with water.** Plants absorb ammonium and nitrate ions through their roots as they take water from the soil. Aquatic producers like green algae can absorb ammonium and nitrate ions from the water around them.
- **Consumers get nitrogen by eating other organisms.** Consumers break down the proteins in their food and use the amino acids to build their own proteins. Consumers release excess nitrogen as nitrogen-rich waste compounds, such as urea, uric acid, and ammonia.

✔ **Decomposers get nitrogen by breaking down dead matter, releasing some nitrogen back to the soil in a process called *ammonification*.** As decomposers break down the proteins in dead things, they may not need all the nitrogen from those proteins for themselves. If the decomposers have excess nitrogen, they will release some of it as ammonia (NH_3) into the soil. Ammonification also occurs when bacteria convert the waste products of animals such as urea or uric acid into ammonia in the soil. Once ammonia enters the soil, it reacts with water to form the ammonium ion (NH_4^+).

✔ **Bacteria change the form of nitrogen as they use molecules for their energy metabolism:**

- *Nitrification* **converts ammonia to nitrite and nitrate ions.** Certain bacteria can get energy by converting ammonia (NH_3) into nitrite (NO_2^-). Other bacteria can get energy by converting the nitrite (NO_2^-) into nitrate (NO_3^-).

- *Denitrification* **converts nitrate ion to nitrite ion and nitrogen gas.** Some bacteria in the soil use nitrate ion (NO_3^-) instead of oxygen to do cellular respiration (see Chapter 8). When these bacteria use nitrate, they convert it into nitrite ion that's released into the soil or into nitrogen gas that's released into the atmosphere.

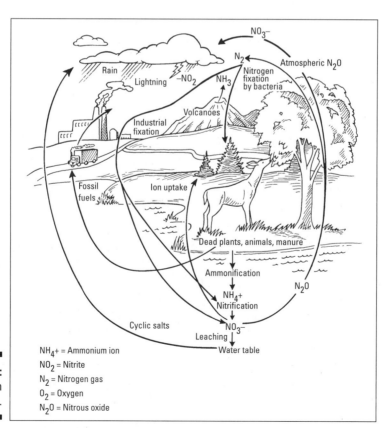

Figure 18-5: The nitrogen cycle.

Interacting with Other Organisms

Organisms are constantly interacting with each other in ecosystems as they work to get the matter and energy they need to survive. Sometimes, these interactions become more than just a temporary association and a long-term relationship may be established.

Living together

When two organisms live together for a significant part of their life cycle, scientists call their relationship *symbiosis*. Some people use the term symbiosis in a nonscientific way to mean a relationship that is positive for both parties. But to scientists, symbiosis doesn't always lead to mutual benefit. So be sure you use the term correctly when you're in a science class!

Plants are part of many fascinating symbioses. Here are a few of my favorites:

- **Several species of acacia trees have a mutualistic relationship with ants.** The trees provide a home and food for the ants, while the ants protect the tree. In Central America, the acacias make large, swollen thorns at the base of the leaves. The ants hollow out these thorns and use them as nests. The trees provide all the food the ants need, producing both a sugary nectar for adult ants in glands, called *nectaries,* at the base of the thorns and making structures high in protein called Beltian bodies that help the ant larvae grow. The ants protect these resources, helping to defend the tree against attack. If a climbing vine tries to wrap itself around the acacia, the ants will chew on the climbing tendrils and free their tree. If grazing insects land on the acacia, the leaf movement brings ants running. The ants will bite and sting any insects that land on the plant, causing the insects to leave and seek other prey.

- **Bromeliads called tank plants that live high in the canopy of the tropical rain forest create mini-ecosystems in the trees.** The tank plants are epiphytes that grow up in the canopy by attaching to rainforest trees. The relationship between the tank plants and the trees is commensal — the tank plants benefit by being able to access sunlight and don't seem to hurt their host tree. The fun part of this symbiosis, though, is that the tank plants form small pond ecosystems up in the tree canopy. The bases of the bromeliad's leaves press tightly together, forming a cup in the center of the plant. This cup fills with water, forming a miniature pond that is home to bacteria, protists, and small animals, such as tree frogs. Although many animals just drop by for a drink or a meal, which isn't a symbiotic relationship with the tank plant, other animals spend their whole lives in the plant. The animals benefit by getting a home and a source of water, and the plant gathers mineral nutrients from the animal wastes — which is very important to a plant that isn't rooted in the soil!

- **Lichens are a mutualistic symbiosis between algae and fungi.** Although lichens aren't plants, they are very plant-like. You've probably seen them growing as green or gold-colored crusts on rocks, or as flat or leafy green patches on trees. The colored part of the lichen comes from the photosynthetic algal partner in the symbiosis. The algae do photosynthesis, making food for themselves and their fungal partner. The fungus wraps around the algae, providing protection as it also absorbs water for the partnership. The lichen symbiosis is incredibly successful, enabling lichens to grow where food and water are very scarce, often making them one of the few living things found in harsh environments.

- **Some plants, such as alder trees and plants in the bean family, form mutualisms with nitrogen-fixing bacteria.** Nitrogen is incredibly important to plants, and it's a mineral that isn't always readily available in the soil. So, some plants take out an insurance policy to make sure they always have a source — they encourage nitrogen-fixing bacteria to move right into their roots and set up house. The plants release food molecules from their roots that attract the bacteria from the soil. The bacteria penetrate into the root cells and even change their shape as they settle into permanent residence. The bacteria enjoy home-cooked meals of sugars from the plant, and as well-behaved guests, they provide the plants with much needed molecules containing nitrogen.

Working things out

The relationships between organisms affect the entire structure of ecosystems.

The loss of an American icon

The forests of Eastern North America were once home to the stately American chestnut tree, which people loved for their beauty, their nuts, and their wood. Then, in the late 19th century, plant material imported into the United States from Asia brought with them an unwelcome hitchhiker — the chestnut blight fungus, *Cryphonectria parasitica*. Although Asian species are resistant to this fungus, the American chestnuts had no defenses. The parasitic fungus decimated the American forests, and by 1950, all that remained of the once glorious American chestnuts were tree stumps and shrubby root sprouts. The trees have struggled on, producing new sprouts that live for a few years until they're attacked and killed by the fungal parasite. Some good news may be on the horizon however. The American Chestnut Foundation is dedicated to bringing back the American chestnut and has made some progress on breeding blight-resistant chestnut trees. And botanists are searching for the genes that make Chinese chestnuts resistant to the fungus, hoping that these genes could be introduced into the American species.

On its website (www.acf.org/index.php), the American Chestnut Foundation describes the reasons behind their efforts to save this native tree.

One especially important type of interaction that plays a role in structuring communities, including those of plants, is competition. *Competition* occurs when two individuals negatively affect each other by reducing access to resources. Resources, such as light, water, space, and minerals, are usually limited within ecosystems, and organisms must compete with each other to get what they need to survive. Competition among plants for resources can lead to reduced plant size and decreased reproduction. Competition can occur between individuals of the same species, called *intraspecific competition*, or between individuals of different species, called *interspecific competition*. It may occur for a short time or for the life of an individual.

Relationships between organisms, including symbiosis, can be positive or negative for the organisms involved:

- **In a *mutualism*, both organisms benefit.** For example, fungi in the soil form partnerships with plant roots. Fungi called mycorrhizae grow on, and even within, the plant roots. The mycorrhizae help the roots absorb water and minerals (especially phosphates) from a wider area within the soil. The mycorrhizae benefit from the relationship because they get some sugar from the plant.

- **In *predation, parasitism*, and *herbivory*, one organism benefits at the expense of the other.**
    - In predation, one organism kills and eats the other.
    - In parasitism, one organism slowly feeds off the other and may eventually lead to its death.
    - In herbivory, grazing animals eat plants, damaging the plants but not necessarily killing them.

- **In *commensalism*, one organism benefits, while the other is not affected.** Epiphytic plants that grow on rainforest trees benefit by gaining access to the sunlight at the top of the forest canopy without negatively impacting their host plant.

- **In *amensalism*, organism has a negative impact on another without being impacted itself.** Amensalism happens when one plant slows the growth of another by shading it from the sun. Some plants engage in a special kind of amensalism called *allelopathy*, where one plant releases a chemical that inhibits the growth of another. For example, plants may release chemicals into the soil that prevent seeds of other plants from sprouting.

Peeking at Powerful Plant Communities

Plant communities are usually the largest and most visible part of an ecosystem, making up a significant portion its biomass.

Biomass is the total mass of all the organisms living in a specific area. The environmental conditions of an ecosystem, such as the amount of rainfall, the intensity of light, the temperature range, and the availability of minerals in the soil determine which species of plants will grow in a particular ecosystem. Because it provides food and homes, the plant community, then has a great impact on the types and distribution of other species in the ecosystem.

Community ecology is the branch of ecology that studies the interactions, abundance, and distributions of particular communities, or mixtures, of organisms. Community ecologists study the behavior of communities almost as if they are an organism, with many interdependent parts.

Exploring biomes

Some communities of organisms cover huge areas of land or exist on multiple continents around the globe.

Biomes are distinct communities that are identified by the major groups of plants and animals that make up the community and the climate conditions to which the community is adapted. Six major types of biomes exist on earth:

- **Freshwater biomes** include ponds, rivers, streams, lakes, and wetlands. Only about 3 percent of the earth's surface is made up of freshwater, but freshwater biomes are home to many different species, including plants, algae, fish, and insects. Wetlands, in particular, have the greatest amount of diversity of any of the freshwater biomes. For details on plants that are adapted to life in the water, check out Chapter 13.

- **Marine biomes** contain saltwater and include the oceans, coral reefs, and estuaries. They cover 75 percent of the earth's surface and are very important to the planet's oxygen and food supply — more than half the photosynthesis that occurs on earth occurs in the ocean.

  Estuaries are areas where saltwater mingles with freshwater. They include familiar places, such as bays, sounds, lagoons, salt marshes, and beaches. Estuaries are important habitat for many different species, including birds, fish, and shellfish. Because they provide habitat for young fish, estuaries are extremely important to the health of fisheries. Unfortunately, estuaries are typically found on the coast, which is also prime real estate for people. As a result, estuaries are being heavily impacted by human development.

- **Desert biomes** receive very low amounts of rainfall and cover approximately 20 percent of the earth's surface. Light intensity may be high, and temperatures can fluctuate from high heat during the day to cold during the night. Desert animals, such as snakes, lizards, birds, and rodents, may be more active at night when temperatures are cooler. Plants that live in deserts have special adaptations, such as the ability to store water or only grow during the rainy season, to help them survive in the low water environment. For details on desert plant adaptations, check out Chapter 13.

- ✔ **Forest biomes** contain many trees or other woody vegetation, cover about 30 percent of the earth's surface and are home to many different plants and animals, including trees, skunks, squirrels, wolves, bears, birds, and wildcats. They're important for global carbon balance because they pull carbon dioxide out of the atmosphere through the process of photosynthesis. Forests are being heavily impacted by human development as humans seek additional land for homes and agriculture or cut down forests for their wood.

    Rain forests are evergreen forests that receive lots of rainfall and are incredibly rich in species diversity. As many as half the world's animals live in rain forests, including gorillas, tree frogs, butterflies, tigers, parrots, and boa constrictors. For more information on rainforest plants, see Chapter 13.

    The *taiga* or *boreal forest* occurs just south of the Arctic tundra. It's dominated by evergreen trees, such as spruce, pine, and fir, but some deciduous trees, such as birches, aspens, and willows, may occur in the wetter areas. Winters in the taiga are long and harsh, and relatively few species of animals live there. Smaller animals, such as birds and rodents, are fairly common, and a few big species, such as, moose and caribou can also be found there.

    The *temperate deciduous forest* is dominated by broad-leafed trees that drop their leaves in the fall and are dormant during the winter. These trees include maples, birch, oak, beeches, basswoods, and hickory trees. During the summer, when the trees are in full leaf, the forest is shaded so that only shade-tolerant plants can thrive on the forest floor. Some species of wildflower avoid this problem by blooming early in the spring before the trees extend their leaves.

    Mountain forests contain some of the largest trees in the world, including coastal redwood trees in California and old growth Douglas fir trees in Oregon and Washington. Most mountain forests experience dry summers, but may receive lots of rainfall in the winter. The trees found in mountain forests are very distinct to the elevation and geographic area. For example, if you were to hike up a mountain in the Sierra Nevadas, you'd begin in forests of Ponderosa pine in the lower elevations, then move through forests of sugar pine, white fir, and Jeffrey pine at intermediate elevations, and finally end up in forests of red firs and mountain hemlock at the higher elevations.

- ✔ **Grassland biomes** are dominated by grasses, but they're also home to many other species, such as birds, zebras, giraffes, lions, buffaloes, termites, rabbits, mice, and hyenas. Grasslands cover about 30 percent of the earth's surface and are typically flat, have few trees, and possess rich soil. Because of these features, people converted many natural grasslands into agricultural use. In North America, for example, huge herds of buffalo once roamed on wild prairies that grew tall grasses and many species of wildflowers. Most of these former prairies now grow cereal crops or are used for cattle ranching.

- **Tundra biomes** are very cold and have very little liquid water. Tundras may be cold enough year-round that they contain permafrost, soil that remains permanently frozen and can prevent penetration by plant roots. Tundras cover about 15 percent of the planet's surface and are found at the poles of the earth as well as at high elevations. The vegetation of Arctic tundras is mostly made up of dwarf shrubs, sedges, grasses, lichens, and mosses. Although tundras may appear treeless, miniature willow and birch trees may survive in some areas. During the growing season, the tundra bursts forth with color from perennial wild flowers. Arctic tundras are also home to animals, such as arctic foxes, caribou, and polar bears, whereas mountain tundras are home to mountain goats, elk, and birds. In both types of tundra, nutrients are typically scarce, and the growing seasons are quite short.

Changing communities through succession

Major changes in an ecosystem, called *disturbances,* can greatly impact a community, even wiping it out completely. But life bounces back and over time, sometimes in as little as a few months, organisms will return to a barren environment and begin to grow, slowly changing it so that other organisms can also live there.

Succession occurs when species of plants and other organisms gradually change their environment, making the environmental conditions favorable for a different set of organisms. Two types of succession occur in communities:

- ***Primary succession* occurs in environments that have new substrate and includes the formation of soil.** For example, when lava cools after a volcanic eruption it forms a bare rock surface. Lichens may colonize tiny cracks in the rock, producing acids and contributing organic matter as they die. These new conditions may allow some species of moss to grow, increasing the amount of organic matter and eventually allowing ferns or seed plants to establish.

- ***Secondary succession* occurs as the composition of organisms changes in environments where soil and vegetation already exist.** Secondary succession typically occurs after natural disturbances like fires, floods, tree-falls, and tropical storms, as well as human-caused deforestation.

After a disturbance, a successional sequence occurs:

1. **The first plants to colonize an area form the pioneer community.**
2. **The successional sequence occurs as each community alters the environment by its growth, then gives way to the next community.**
3. **The final, stable community that forms at the end of the successional sequence is the climax community.**
4. **The climax community remains in place until a new disturbance restarts the cycle of succession.**

During a successional sequence, changes occur to both the community and the ecosystem. Communities change as the types of species change. As the community approaches climax, the diversity of species typically increases as does the number of perennial and woody plant species. In the ecosystem, light becomes less available, but soil nutrition and structure improves, and the soil becomes deeper and more able to hold moisture. The total biomass of the ecosystem increases, as do the amounts of photosynthesis and respiration.

The total amount of biomass produced by the plants of an ecosystem in a given amount of time is called its *primary productivity*. Primary productivity is a measure of the food production capability of an ecosystem and also indicates its ability to function as a carbon sink.

Fire!

Wildland fires are destructive and can be dangerous to human habitation. However, fires are a natural disturbance that occur periodically in ecosystems as a result of lightning strikes or human actions.

Fire ecology studies the role of fire in ecosystems. Fire ecologists are interested in origins and spread of fire and its impacts on ecosystems, including its possible role in maintaining ecosystem health.

Fires have many effects on ecosystems:

- **Fires have a mosaic effect on ecosystems, which results in increased diversity of vegetation.** After a fire, some areas will be completely burned, while others are only partially burned. In each area, different pioneer communities will take hold, creating diverse patches within the ecosystem.
- **Fire can be beneficial or detrimental to soil.** Ash and charcoal increase the nutrient content in soil. If a fire is very hot, however, soil particles change so that they repel water. As water runs off this fire damaged soil, it carries the soil particles with it, speeding up the process of *erosion*.
- **Fire kills some individual animals, but may allow certain populations to thrive.** Small animals that can't successfully flee the fire die, but fire typically increases food availability for scavengers and prey visibility for predators.

- **Fire and plants mutually impact each other.** The type of vegetation affects how quickly fire spreads and how hot it burns. Likewise, the hotter the fire, the greater the destruction to the plant community. By creating a disturbance and increasing the availability of light and soil nutrients, fire can actually improve growing conditions for some of plants. Plant species may have adaptations that help them survive fire, and some actually require fire in order to reproduce! For example, giant sequoias can produce bark that's up to two feet thick to help them survive fire and lodgepole pines require fire to release their seeds from their cones.

Because of the beneficial effects of fire on ecosystems, some people recommend *prescribed burns* as part of the management plan for wildlands. Prescribed burns are either fires that are intentionally set or natural fires that are allowed to burn

Human disturbances to ecosystems

People rely upon the health of natural ecosystems for food, clean water, clean air, and places to live. Just like all organisms, humans take what they need from their environment and release wastes. However, people have developed tools and technology that enable them to dramatically transform human environments and extract natural resources at unprecedented rates. When you combine the human ability to transform the environment with the size and rate of growth of the human population, human impacts on natural ecosystems become dangerously large.

Natural systems are valued by humans for their beauty and the products the produce. Unfortunately, many human actions damage natural ecosystems:

- The burning of fossil fuels is contributing to the greenhouse effect and global warming, causing the temperature of the earth to increase. Global warming has the potential to change the weather patterns of the earth, which will affect agriculture and the habitats of many species.

- Another effect of the burning of fossil fuels is an increase in air pollution and the production of *acid rain* and *acid fog*. The burning of coal and fossil fuels produces sulfur dioxide and nitrogen oxides. Water vapor in the atmosphere mixes with these chemicals, producing sulfuric acid and nitric acid. When precipitation falls, the drops of water (or flakes of snow) take the acids with them down to soil, lakes, and rivers, killing organisms in these ecosytems.

- Certain chemicals produced by people, such as the chlorofluorocarbons (CFCs) used in aerosol sprays and older refrigeration units (made before 1996), evaporate into the atmosphere and erode the *ozone layer*. The ozone layer is important because it helps protect organisms from ultraviolet rays that can damage DNA.

- ✓ Chemical pollutants become concentrated in organisms at the top of food chains, harming those organisms and people who eat them. Pollutants may be absorbed at low concentrations by organisms at the bottom of the food chain. But if these organisms are eaten in large numbers by something higher on the food chain, the concentration of the pollutant is increased by this *bioaccumulation*. High levels of mercury in tuna fish are an example of bioaccumulation at work.

- ✓ Lawn-care or cleaning products that contain phosphates and nitrates add these nutrients to natural waterways, causing *eutrophication*. Eutrophication leads to massive overgrowth of algae, called *algal blooms*. When the algae die, decomposers break down the dead algae, using up oxygen as they do cellular respiration. Eventually, so much oxygen is removed from the pond, lake, stream, or river that invertebrate animals and fish die.

- ✓ Another way that humans are disrupting nature is by removing habitats altogether. The rain forests that form the belt of the world all around the equator are huge carbon sinks, provide large amounts of oxygen to the atmosphere, and harbor many different species. As humans destroy rainforests and other habitats, they reduce the diversity of living organisms, called *biodiversity,* reducing the health of ecosystems and the availability of organisms for human uses.

Conservation biology studies the effects of humans on the environment and the importance of biodiversity. Conservation biologists work to understand and prevent future extinctions of species.

Chapter 19

Altering Plants by Using Biotechnology

In This Chapter
▶ Defining genetically modified organisms (GMOs)
▶ Looking at the genetic modification process
▶ Debating the pros and cons of GMOs

Scientists produce genetically modified organisms by introducing genes for desired traits into useful organisms, such as crop plants, farm animals, and bacteria. The engineered organisms may be easier to grow, or they may produce valuable medicines. However, many people, including scientists, are worried about the risks of genetic engineering. GMOs may contaminate other crops or adversely affect natural ecosystems. This chapter offers an introduction to GMOs, describes a method for transforming plants, and then looks at the benefits and concerns surrounding this technology.

Oh, No! GMOs!

You've probably heard a lot in the news about genetically modified organisms (GMOs). Agricultural companies are making them. Some people are afraid of them. The European Union has very tough rules about allowing them into the European food supply.

Genetically modified organisms, which are also sometimes called *transgenic organisms,* are organisms whose genetic code has been changed using recombinant DNA technology. *Recombinant DNA technology* refers to the tools that scientists use for manipulating DNA — in other words, for taking genes out of one organism and putting them into another.

Recombinant literally means having DNA from two different organisms, so you can think of something that's recombinant as having a combination of DNA from at least two sources.

When scientists use recombinant DNA technology to change the genes of organisms, they call it *genetic engineering*. *Genetic engineering* is the manipulation of DNA using recombinant DNA technology to add, delete, or change a trait in an organism.

Scientists have many reasons for using genetic engineering to alter the traits of organisms:

- Animals used in agriculture may be genetically engineered to make them grow faster, produce more milk, or make leaner meat.
- Crop plants may be genetically engineered to be more resistant to diseases, insects, or weed killer or to make additional vitamins.
- Bacteria and viruses may be altered to make them less harmful to people so that they can be used in vaccines.
- Bacteria can be genetically engineered with human genes so that they make human proteins that can be used as medicines.

At least some of these goals seem beneficial to humanity. Still, people have many questions and concerns about the safety of these efforts. Genetic engineering allows agricultural scientists to do almost immediately what it used to take traditional breeding programs decades to achieve. People wonder if these scientists are moving too fast, if they've really looked at all the possible consequences of creating GMOs.

Engineering Plants

Several characteristics of plants make it easy for scientists to genetically modify them:

- Plant cells and tissues can be grown in media in the lab.
- Pieces of DNA can be introduced into plant cells:
    - A bacterium called *Agrobacterium tumefaciens* (crown gall bacteria) infects plants, naturally introducing a piece of DNA into plant cells in the process.
    - Scientists use *gene guns* to shoot small pieces of metal that are coated with DNA into plant cells.
- Scientists can use hormones to induce clumps of plant cells to differentiate into complete plants.

Culturing plant tissues

Tissue culture is the process of growing tissues on growth medium in a lab. Figure 19-1 illustrates the process of growing plant cells in culture:

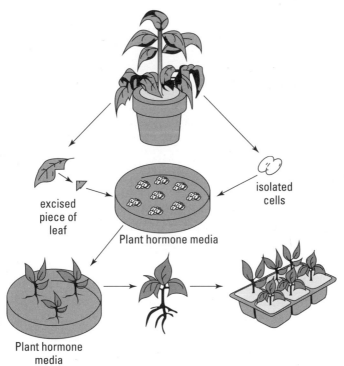

Figure 19-1: Plant tissue culture.

Overview of the Tissue Culture Process

- Plant cells are taken from any part of the plant, although tissues taken near meristems usually grow the best.
- The pieces of tissue are placed on special growth media containing hormones that favor the growth of plant cells.
- The cells begin to grow and form an undifferentiated blob of cells called a *callus*.
- Hormones are added to the medium to cause the callus to grow roots and shoots, thus developing into a complete plant.

Transforming plants with Agrobacterium

When *Agrobacterium* attacks plants, it transfers a small piece of DNA, called the *Ti plasmid*, into the plant cells.

Genes from the Ti plasmid insert themselves into the plant DNA, causing the plant cells to divide repeatedly, producing a lump of tissue called a gall at the crown of the plant where it enters the soil. The rapid cell division in the plant is very similar to the rapid cell division caused by cancer in animals.

Cutting DNA with restriction enzymes

Scientists use the unique talents of *Agrobacterium* to modify plant cells in order to insert specific genes into plant cells. To do so, the scientists first have to modify the Ti plasmid to carry the gene they want to put into plant cells, which is called the *gene of interest*. Scientists use enzymes called *restriction enzymes* to cut the DNA of the Ti plasmid and insert their gene. Restriction enzymes cut DNA at very specific sequences, and each restriction enzyme cuts at a unique sequence.

Figure 19-2 shows how restriction enzymes can be used to cut open the Ti plasmid, which is a circular piece of DNA. Once the Ti plasmid is cut open, scientists can add their gene to the plasmid.

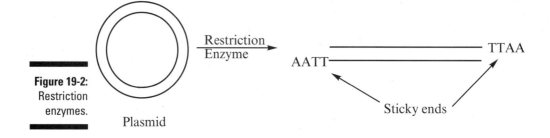

Figure 19-2: Restriction enzymes.

Inserting a gene into the Ti plasmid

Restriction enzymes make the process of cutting and combining pieces of DNA easy. To put their gene into the Ti plasmid, scientists follow these steps (se Figure 19-3):

Chapter 19: Altering Plants by Using Biotechnology 315

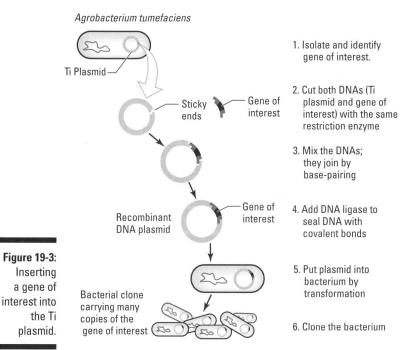

Figure 19-3: Inserting a gene of interest into the Ti plasmid.

1. **Choose a restriction enzyme that cuts DNA asymmetrically.**

    Restriction enzymes that cut the DNA backbone asymmetrically leave a piece of single-stranded DNA hanging off each end. These pieces of single-stranded DNA, called *sticky ends,* are complementary to other pieces of DNA cut with the same restriction enzyme and can form hydrogen bonds with them (see Chapter 2). For example, the sticky ends shown in Figure 19-2 have the sequences 5´AATT3´ and 3´TTAA5´. A and T are complementary base pairs, so these ends could form hydrogen bonds and thus stick to each other.

2. **Cut the gene of interest and Ti plasmids with the same restriction enzyme.**

    If you cut the Ti plasmid and the gene of interest with the same restriction enzyme, all the DNA fragments will have the same sticky ends.

3. **Combine the gene of interest and Ti plasmids.**

    The two types of DNA have the same sticky ends, so some pieces of plasmid DNA and the gene of interest will stick together. Thus, some plasmids will end up with a human gene inserted into the plasmid.

4. **Use an enzyme to seal the Ti plasmid closed again.**

    An enzyme called *DNA ligase* will form covalent bonds at the cut sites in the DNA, sealing together any pieces of DNA that combined together.

5. Return the modified Ti plasmid to *Agrobacterium*.

 When the bacteria and modified Ti plasmids are mixed together in a test tube, some bacteria will pick up the plasmid and move it into their cytoplasm. Scientists improve the chances of the bacteria picking up the plasmid by shocking them with alternating treatments of hot and cold temperatures.

 6. Grow the bacteria and identify which ones picked up the plasmid.

 When scientists modify the plasmid to add the gene of interest, they also add a gene to help them identify which bacteria have taken up the plasmid. Often they add a gene that will make the bacteria resistant to antibiotics. Then when they grow the bacteria, they add the antibiotic to the growth medium. Only bacteria that took up the plasmid, gaining the gene of interest and the antibiotic resistance gene, will be able to grow.

Putting the Ti plasmid into plant cells

Figure 19-4 shows how scientists use *Agrobacterium* to create transgenic plants.

 1. Scientists punch out disks of leaf tissue from a plant and then combine the leaf disks with modified *Agrobacterium* that carries the gene of interest.

 2. The bacteria infect the plant cells, transferring the Ti plasmid into the cells.

 3. Genes move from the Ti plasmid into the DNA of the plant cells.

 4. Scientists identify which plant cells have gained the gene of interest by looking for a chemical marker.

 5. Scientists grow the modified plant cells in tissue culture.

 6. Scientists use hormones to trigger the plant tissues to differentiate into new plants.

 For more on which hormones would trigger root and shoot development, see Chapter 10.

The new plants are geneticlly modified organisms because they contain the gene of interest inserted by the scientists. The plants contain this gene in every cell and will pass it on to their offspring when they reproduce.

Chapter 19: Altering Plants by Using Biotechnology 317

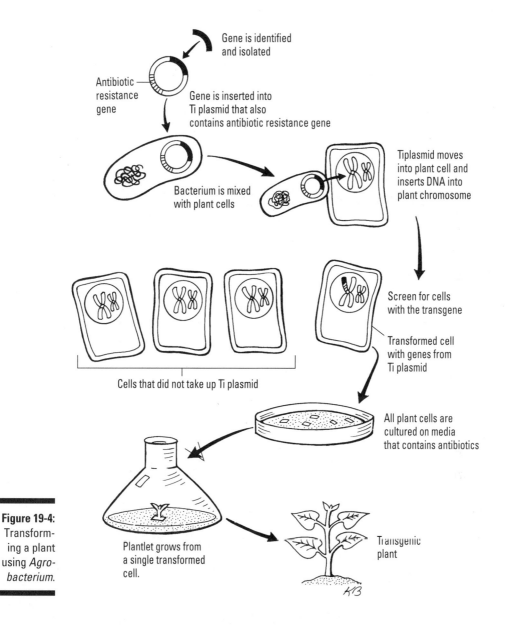

Figure 19-4: Transforming a plant using *Agrobacterium*.

Trying to Build a Better World

People around the world are suffering from malnutrition. And, things could get a whole lot worse as the human population continues to grow, perhaps adding 100 million people per year for the next 30 years. Some scientists believe that they must use genetic engineering to improve agricultural crops to meet these needs and find ways of growing more crops, faster.

Scientists are using genetic engineering to create new varieties of crop plants that are disease and pest resistant, better able to survive environmental extremes like cold and drought, more efficient at drawing nutrients from the environment, and more nutritious for people to eat.

Saving lives with golden rice

Vitamin A deficiency is linked to the death of 1.15 million children a year and causes blindness in many more. Scientists are trying to help these children by using genetic engineering to improve the nutritional content of white rice, which is a staple food in many countries.

Biofortification is the creation of plants that make or accumulate additional nutrients.

If people eat provitamin A, or beta-carotene, it will convert to vitamin A in the body. Rice plants actually make provitamin A in their leaves, but not in the grain, which is what people eat. And people in rural areas of Asia and Africa often prefer white rice, which is especially nutrient-deficient, to the more nutrient-rich brown. However, scientists figured out that only two genes needed to be added to the rice plants to turn on provitamin A production in the grain. They used genetic engineering to add those two genes and created a strain of rice called *golden rice*. The grains of golden rice are yellow instead of white because they contain provitamin A, or beta-carotene, which is yellow-orange in color.

The Golden Rice Humanitarian Board is currently working to distribute golden rice varieties free of charge to farmers in countries where vitamin A deficiency is a problem. However, the situation is a complex one due to cultural, economic, and political factors.

Stopping crop pests with "Bt"

The larvae of moths and butterflies *(lepidopterans)* can do significant damage to crop plants. For example, the larva of the corn borer moth causes damage to corn leaves, stalks, and ears, reducing the yield of corn crops. One way to control lepidopterans is by using a toxin produced by the bacterium *Bacillus thuringiensis,* nicknamed *Bt.* The toxin produced by *Bt* targets the gut of lepidopterans, causing them to stop feeding and die. Many organic farmers use *Bt* to control lepidopterans by spraying the bacterium on their crops. When the larvae eat the toxin from the bacterium, they die.

Scientists have also used genetic engineering to make corn plants that make the bacterial toxin all by themselves. They removed the gene that's the blueprint for the toxin from the bacteria and then introduced it into corn plants, creating *transgenic Bt corn*. The corn plants use the bacterial gene to make the toxin. If larvae graze on these transgenic plants, they die. Farmers don't have to use pesticides or spray *Bt* to protect their crops. Scientists have also introduced the gene from *Bt* into cotton, to protect cotton plants from a pest called the bollworm.

Unfortunately, the larvae of crop pests may not be the only insects that are affected by the transgenic plants. Corn is wind-pollinated, and when the plants are making pollen, the golden pollen grains — containing the *Bt* toxin — get blown everywhere. Scientists at Cornell University sprinkled *Bt* pollen on milkweed and fed it to the larvae of monarch butterflies. The larvae got sick and died. Milkweed and monarch butterfly larvae are often found at the edges of cornfields. Other scientists have raised the possibility that pollen and plant material from transgenic corn may harm aquatic ecosystems when it blows into nearby streams. If larvae of stream insects are killed by the toxin, then the entire food web may be affected and organisms like fish won't have enough food. So, while transgenic *Bt* plants may solve some problems, they may cause others.

Resisting weed killers

Another problem that farmers face is the problem of weeds. The safest solution in terms of the environment is to pull them by hand, but this practice is extremely time-consuming. So, many farmers turn to weedkillers as a way to kill plants that are competing with crop plants for nutrients. One of the most common weedkillers in use is the chemical *glyphosate,* which is the active ingredient in weedkillers like Roundup. Some farmers will spray an entire field with glyphosate to kill all the weeds before planting the field with soybean or corn. Spraying a field to kill weeds is much easier than plowing the weeds into the soil.

The company Monsanto is marketing a product that's designed to make the use of Roundup even easier. Monsanto used genetic engineering to introduce a gene that makes crop plants able to survive the application of Roundup. Roundup Ready soybean is a variety of soybean plants that Monsanto genetically engineered to survive Roundup.

If farmers plant Roundup Ready crops, they can spray their fields with Roundup, killing all the weeds and leaving the crop plants behind. Monsanto has also produced Roundup Ready corn, cotton, and canola. The use of these crops is popular with some farmers because growing them is much easier.

The use of Roundup Ready crops is very controversial. People who support these crops argue that glyphosate is less toxic than other weedkillers and that the use of Roundup Ready crops allows farmers to grow larger crops with less effort. But people also argue against the use of the crops because it encourages the use of greater concentrations of Roundup and the use of weedkillers in general, and Roundup is toxic to plants and fish in the environment.

The use of Roundup by farmers for the past 30 years may soon make the argument a thing of the past. By using Roundup over and over, farmers have consistently killed the susceptible weeds, leaving the most resistant weeds to survive. Farmers are now having to deal with *superweeds* that Roundup can't kill. Unfortunately, many farmers are responding to this new problem by using more toxic weedkillers.

Making drugs and commercial enzymes with pharma crops

Recently, companies have begun engineering crop plants like corn to produce drugs and vaccines for use in medicine and enzymes for use in commercial processes. Scientists create *pharma crops* by introducing a gene for a drug or an enzyme into a crop plant like corn, rice, barley, or soybean. The plants are grown in open fields on farms. After harvest, the desired product is purified from the plant material.

The companies that are producing these products claim that they'll lower the production cost of pharmaceuticals, making drugs available at cheaper prices and saving lives. They also say that these crops could provide a new commercial venture for farmers in depressed rural areas, thus helping generate revenue and save farms.

However, activists and many scientists express high levels of concern over these products. Pharma crops are being grown in open fields. Pollen released from these crops may spread to other crop plants or to wild plants, transferring the genes for these products into other plants. People and animals may then end up eating foods that contain drugs, hormones, or commercial enzymes.

Let's Talk, Talk, Talk about It: Pros and Cons of Genetic Engineering

The use of genetic engineering to alter living things is controversial for many reasons: People have questions about the benefits, safety, environmental impacts, financial rewards, and even the ethics of these practices. And yet the use of GMOs in agriculture is increasing to the point that it's getting hard to find foods, especially those that contain corn and soy, that don't have a genetically engineered component. Many scientists have entered the debate about these products, examining the benefits and expressing concerns.

The pros of GMOs:

- ✔ Crop plants that resist weedkillers are easier to grow, which can increase crop yields and profits for the farmer.
- ✔ Crop plants that resist agricultural pests may reduce the need for pesticide use, protecting the environment from their effects, decreasing costs for the farmer and producing higher yields.
- ✔ If crop plants or farm animals raised for human consumption are given genes to improve their nutrition, people could be healthier.
- ✔ If farm animals raised for human consumption are given genes to increase their yield of meat, eggs, and milk, then more food may be available for the growing human population, and these greater yields may also increase profits for farmers.
- ✔ Medicines and vaccines produced in pharma crops may be cheaper and more available to the human population.
- ✔ If the human population continues to increase without changes to current agricultural practices, food may become scarce leading to increased starvation.

The no's of GMOs:

- ✔ Crop plants that contain genes for toxins that kill insects may harm insects in nearby ecosystems.
- ✔ Engineered crops may pollinate other crops or wild relatives, transferring the introduced genes into the food supply in an uncontrolled fashion. Once these genes have been transferred, it's almost impossible to track down all the affected plants and remove them from the human food supply. Many potential problems could arise from the transfer of genes into wild plants.
- ✔ The profits of GMOs mostly go to the companies that develop them, not to farmers or consumers. Companies patent the engineered organisms they develop and carefully control the seed supply. Farmers must purchase new seed each year and aren't allowed to save seed from a crop as that is considered copyright infringement. If a small farmer can't afford to buy the engineered crop, he must compete with farmers who can, potentially making it harder for small farms to stay in business.
- ✔ Scientists question whether GMOs are necessary and if they represent the best way to solve agricultural problems. GMOs may introduce as many agricultural problems as they solve. And typically, agricultural biotechnology doesn't encourage sustainable farming practices because, in order to be profitable, companies develop products that must be purchased over and over again.

- Animals that are engineered to produce more milk, eggs, or meat may be at greater risk for health problems.

- Genetic modification of foods may introduce allergens into foods, and labeling may not be sufficient to protect the consumer. For example, a number of years ago, brazil nut protein was added to soybeans in Europe, resulting in allergic reactions in people with nut allergies who ate products containing the modified soybeans.

- Very few rigorous scientific studies have investigated the safety of GMOs, and the true risks remain unclear. More research by independent scientists, as opposed to those employed by the large agricultural corporations, needs to be done.

Many questions remain to be answered, and yet more and more GMOs are entering the human food supply. The big agricultural firms that produce these products make a great deal of money and are active supporters of this technology. Personally, I think we are moving too fast, but I don't think we should "just say no to GMO" and throw out the good with the bad. If GMOs can deliver vitamins to malnourished children or vaccines to those at risk for disease, then the benefits of those products outweigh the risks. However, I'm not personally in favor of genetic modifications that encourage unsound environmental practices, large scale farming of limited crop species, or cruelty to animals. We need to continue to discuss these issues as a society and weigh the pros and cons of each situation. In these debates, some important questions to ask are "What are the risks versus benefits?", "Who benefits?", and "Who is put at risk?"

Chapter 20
Thriving on Plants in Everyday Life

In This Chapter
▶ Growing plants for food
▶ Making paper, clothing, and biofuels
▶ Exploring plant chemistry

*P*eople have been cultivating plants for more than 12,000 years, which just goes to show you how important plants are to people. As the human population grows, new methods are needed to improve food production. Plants are also important to people for building material and useful products, such as paper and clothing. However, conventional techniques for making these products can harm the environment. This chapter looks at some of the ways that people use plants and explores how people are looking for new, more sustainable ways to support the human population.

Feeding a Hungry World

What did you eat today? How many foods do you normally consume that come from plants? Chances are, much of what you eat comes from plants and, in particular, from seed plants.

Eighty percent of the calories that humans eat come from six seed plants: wheat, rice, corn, potatoes, cassava, and sweet potatoes. These plants have been staples in the human diet for thousands of years, ever since people first started farming.

Exploring the origins of agriculture

Early humans were hunter-gatherers, moving around as they hunted wild animals and gathered wild plants from their environment.

About 12,000 years ago, humans shifted from hunting and gathering to agriculture.

Once people became farmers, they settled into communities that stayed in one place. The structure of these communities became the basis for modern civilization as it exists today. Some of the earliest agricultural sites are located in the *Fertile Crescent*, a fertile area in the Middle East, southeast Asia, and China that follows the Nile, Tigris, and Euphrates rivers, forming a broad semi-circle at the Northern end of the Red Sea. The ancient civilizations of Egypt, Phoenicia, Assyria, and Babylonia farmed in this area, which today includes Egypt, Israel, Lebanon, and portions of Jordan, Syria, Iraq, Turkey, and Iran.

The first plants to be cultivated were probably wild plants that people had previously gathered. Instead of collecting wild plants, people began to cultivate them. Figure 20-1 shows a world map marked with the kinds of plants that were first grown by early farming communities around the world.

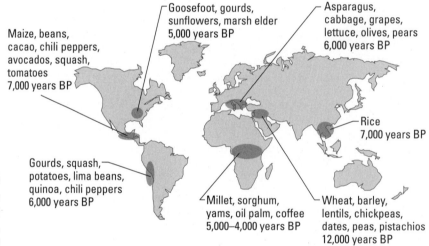

Figure 20-1: Origins of agriculture.

Some of the world's most important modern crops have been farmed since these early days:

- People first cultivated grains, such as wheat and barley, in the Fertile Crescent around 12,000 years BP (before present).
- The ancient Chinese started growing rice around 10,000 BP.
- Central Americans began farming teotsinte, the ancestor of corn, around 9000 BP. They also began growing beans and cacao, which is used to make chocolate, at least as early as 7000 BP. (Thank you, ancient Central Americans!)
- Mediterranean people began cultivating olives around 6,000 BP.
- South American people were growing potatoes 6,000 BP.
- Central Africans started farming coffee around 5,000 BP.

As people domesticated their local wild plants, they began to change them. People chose the individual plants that had traits they liked best and bred them to create the next generation. Over time, this constant selection for certain traits led to crop plants that were significantly different from their wild cousins.

Keeping up with human population growth

Almost seven billion people live on planet earth today, and the number keeps increasing. Around 925 million of those people don't have enough to eat. For the children of the world, these aren't just numbers: 10.9 million children under age five die each year. Sixty percent of those deaths are linked to malnutrition. And the number of people on earth keeps increasing — estimates vary, but a recent estimate by the United Nations predicts that almost nine billion people will inhabit the earth in the year 2050.

And while the number of people keeps growing, the earth is staying the same size. More importantly, the amount of land that's usable as farmland remains the same size or may actually be decreasing due to soil degradation and desertification.

Only about 3.5 percent of the earth's surface is suitable for farming, and most of that is already being farmed. Humans use about 40 percent of the world's land surface, dedicating about 37 percent to agriculture. Of the area used for agriculture, one-third is used for food crops, and the rest is used for feed crops for animals or land for grazing.

If you pushed all the crop-growing land of the world together, it would be about the size of South America. If you put all the land used for grazing together, it would be the size of Africa. (Urban areas occupy an area about one-third the size of Europe.)

To feed an additional two billion people by 2050, governments, scientists, and farmers are going to have to figure out how to get more food out of the land that's already being farmed.

People are already working on ways to come up with more food:

- **Eat less meat.** You have to feed a cow about 10-15 pounds of grain to make 1 pound of beef. Even though beef has more calories per weight than grains, you can still satisfy the caloric needs of a lot more people with 10-15 pounds of grain than you can with 1 pound of beef.

- **Produce a higher yield of crops from the land already in use.** Scientists and farmers have already demonstrated their abilities to increase yields through improvements in crop varieties, irrigation, fertilization, and pest control. (See the sidebar "The green revolution" for details.) During the years 1987 to 1996, for example, crop yields of soybeans and wheat in the United States doubled, and those of corn tripled. One way that scientists are trying to keep this trend going is by using genetic engineering to create genetically modified crop plants (see Chapter 19).

✔ **Find more *arable land,* which is land that's suitable for farming.**
Unfortunately, even though the best land is already in use for farming, demand is so high that people are attempting to convert other land to farming, even if it's unlikely to be very successful. One of the biggest issues is the loss of forests, called *deforestation,* as people cut down trees to try and convert the land for farming.

Cutting down forests for farming has many negative outcomes that outweigh the positives:

- Forest land may not yield crops for very long, especially in the rainforest, where soils are shallow and nutrient-poor. Unfortunately, people do burn down parts of the rainforest in order to try and grow food. When their efforts fail a couple of years later, they move on and burn another piece.

- Forests are important *carbon sinks:* Forests remove carbon dioxide from the atmosphere through photosynthesis and then use that carbon dioxide to build the mass of the trees. If forests are cut down or burned, the stored carbon in the trees can get released back to the atmosphere as carbon dioxide, increasing the global temperature.

- Forests are home to great diversity of plants and animals. Not only are forests necessary to other kinds of life, they also harbor a great deal of the planet's biodiversity. Many useful products come from forests, and many more might be there waiting to be discovered.

The green revolution

In the 1940s, the world's population was growing rapidly at a faster pace than agriculture was producing food. Realizing that something needed to be done to meet the growing demand, private foundations and national governments from resource-rich nations started programs to improve farming in developing nations. For example, the Rockefeller Foundation in the United States funded a program to improve wheat yields in Mexico. The program, led by Norman Borlaug, spent almost 20 years using plant breeding to produce dwarf wheat varieties that gave a high yield and were resistant to plant pests and diseases.

In the 1960s, Borlaug was part of a group that brought the high-yield wheat to India and Pakistan and taught the farmers how to grow it to get the best results. In the 1980s, China adopted these improved agricultural varieties and practices, resulting in its growth into one of the leading food producing countries in the world. The improvements in food production as a result of the agricultural advances spearheaded by Borlaug were so dramatic that in 1968 the USAID director called it a "green revolution." Borlaug is credited with being the father of this revolution and with helping to save over a billion people from starvation. For his contributions to world peace by increasing food supplies, Borlaug received the Nobel Peace Prize in 1970.

Making Use of Plant Products

People grow plants for a variety of products in addition to food crops. Plants provide building materials, paper, fabrics, spices, dyes, medicines, and fuels, as well as beverages like coffee, tea, and cocoa.

Building homes

Wood, from both gymnosperms and angiosperms, is the most widely used construction material in the world. The demand for timber for construction increases pressure on the world's forests. Resource-rich nations, such as those in North America and Europe, have developed sustainable practices to meet their demand for timber. (*Sustainable* practices are those can be maintained over time with minimal long-term effects on the environment.)

Thus, even though North American and European countries are among the global leaders in timber harvest, their remaining forests aren't being impacted as much as forests in other countries. Europe actually experienced gains in the size of their forests since 1990, and countries in North America experienced small losses. Both of these regions compensate for timber harvest by replanting trees and contain large areas dedicated to forest plantations.

Unfortunately, other areas of the world are not currently harvesting their timber sustainably. In particular, South America and Africa are losing their forests at a staggering rate. They're cutting their trees as fast as European countries, but they don't have as many forest plantations — the trees they cut are often from native forests, leading to a loss in biodiversity.

The amount of wood being used for construction can be determined by looking at the amount of sawnwood being produced.

Globally, European countries produce the most sawnwood, with North America coming in second. Sawnwood production in Asia and South America are both on the rise, where it is partially based on plantation grown timber.

In addition to raising trees for timber on forest plantations, another way to increase sustainability is to use *recycled wood,* which is also called *reclaimed wood.* Up to 40 percent of the wood from demolished buildings can be reused in new construction.

Green construction, which is also called *green building or sustainable building,* is construction that maximizes efficient water, energy, and resource use and reduces waste, pollution, and environmental degradation. Green construction uses materials that are harvested sustainably, such as trees from sustainable plantations or rapidly growing materials such as bamboo.

Making paper

About 42 percent of the world's commercial wood is used to produce the paper people use for writing, reading, wrapping, packaging, and wiping. Industrialized nations, which represent 20 percent of the world's population, use 87 percent of the world's printing and writing papers.

In order to make paper from wood, the cellulose fibers in the wood must be separated from the lignin that holds them together.

The paper-making process

Figure 20-2 shows the traditional process of making paper from trees:

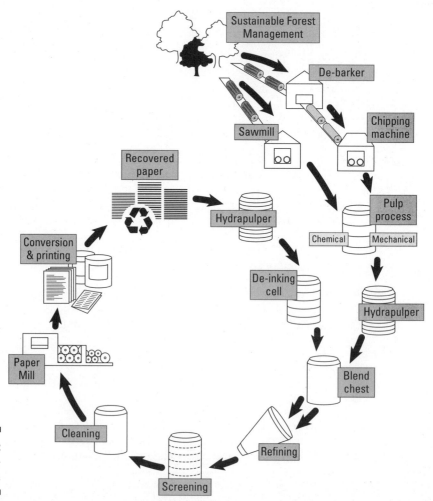

Figure 20-2: Paper manufacturing.

1. **The wood is prepared for pulping by removing the bark and cutting the wood into smaller wood chips by using a machine called a *chipping machine*.**
2. **Wood is turned into pulp.**
3. **The wood pulp is combined with water in a hydrapulper, turning it into a thick pasty substance.**

   Recovered paper can also be used to make recycled paper. The paper is added to a hydrapulper and staples and other contaminants are removed by screening. The ink separates from the paper and is drained away in a process called *de-inking*.

4. **The paste from hydrapulpers passes into the blend chest, and chemicals and dyes are added to create the desired characteristics of the finished paper.**
5. **The paper paste from the hydrapulpers goes through refining as it is passed through a series of revolving metal disks to make it more flexible.**
6. **The paste is then passed through screens and spun rapidly in order to remove dirt and other undesirable materials during the process of screening and cleaning.**
7. **The paste is diluted with water as it enters the wet end of the papermaking machine where it is rolled into sheets and then cut.**

Environmental issues

People today use a lot of paper and that creates environmental problems:

- **Conventional paper bleaching techniques release harmful chemicals into the environment.** In order to create high-quality white paper for printing, paper pulp must be bleached, which releases harmful chemicals into the environment. Paper mills that use conventional chlorine bleaching generate chlorinated organic compounds, such as dioxin, releasing them into the water near the paper mill. These chemicals can remain in the environment for long periods of time, harming other organisms and eventually finding their way to people where they are linked to cancer and lung disease.

- **Most of the world's paper supply comes from forest-harvested trees rather than from tree farms.** The growing demand for paper is increasing the rate at which deforestation occurs, leading to loss of biodiversity and environmental degradation.

- **Pulp and paper mills use large amounts of water and produce greenhouse gases.** High water usage makes water less available for ecosystems as well as for agriculture and other human uses. The production of greenhouse gases contributes to global warming.

As the human population increases and more nations become industrialized, the demand for paper products goes up. Paper production from the 1950s until today has increased by over 400 percent! And from now until 2020, it's projected to increase another 20 percent.

What you can do:

- **Reduce your paper use.** If you can avoid printing something, that's a plus for the environment.
- **Recycle your paper.** You can use waste paper to make new paper, saving trees and lessening the amount of deforestation.
- **Buy paper products made from recycled paper.** By increasing the demand for recycled paper products, you will encourage the paper industry to make paper from recycled paper instead of from timber.
- **Encourage the use of more environmentally friendly materials in the paper industry.** Safer bleaching agents are available and used in countries like Sweden. Also, paper doesn't need to be made from wood. Before people made paper from trees, they made paper from rags. Paper can be made from sawdust, scrap lumber, fabric, even kangaroo poo. (Gotta hand it to the Australians for coming up with that one!) Let your legislators know that you care about pollution and put pressure on the paper industry to change to greener methods of paper production.

Wearing cotton

Cotton provides almost half of the fiber that's used to make clothing and other textiles. It's planted on about 2.4 percent of the world's cropland spread over 90 countries. All together, these countries produce about 20 million tons of cotton per year.

Unfortunately, cotton requires large amounts of water and pesticides in order to be a productive crop, which leads to negative impacts on surrounding ecosystems and the people who live there. The *Sustainable Cotton Project*, located in California, is working to change the way people grow cotton. This project brings farmers, manufacturers, and consumers together to develop less harmful ways of growing cotton and raising awareness about the negative impacts of traditional cotton farming.

Fueling the future

The fastest growing sector in agriculture is in plants that can be used to generate biofuels. *Biofuels* are fuels made from organic matter such as plants, animals, or their byproducts from industrial processes. As fossil fuels, such as coal, oil, and gas, become less available, people are searching for alternative sources of energy. *(Fossil fuels* are energy-rich hydrocarbon deposits formed in the earth from plant remains during a previous geologic age.)

One solution to the energy crisis is to use plant material from plants that are alive today to make fuels, such as ethanol or biodiesel.

Making ethanol

Figure 20-3 shows how cellulose-rich plant material can be used to make ethanol:

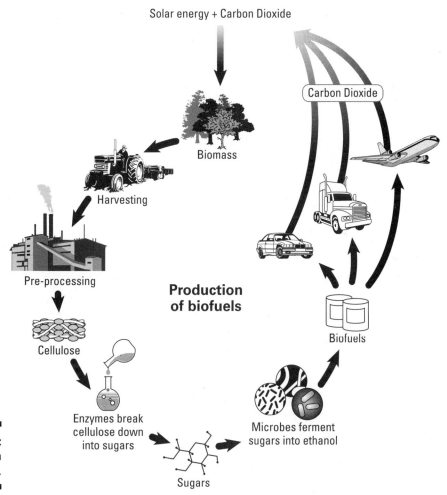

Figure 20-3: Production of biofuels.

1. **Plant material, such as corn, is harvested and ground, or milled, into small pieces, producing a corn mash.**

   Dry milling grinds corn into a flour or meal. *Wet milling* grinds the corn in water so that a corn slurry is produced.

2. **Enzymes are added to the corn mash to break the large carbohydrates, such as cellulose and starch into smaller sugars.**

3. **Yeast is added to the corn sugar.**

   The yeast performs *fermentation,* a process that breaks down the sugar, producing ethanol and carbon dioxide.

Using biodiesel

Biodiesel is diesel fuel made from plant products. Biodiesel can be made from plant oils, including waste oils from cooking processes. The plant oils are converted to biodiesels using lye and an alcohol like methanol.

For some, biofuels from crop plants represent an attractive alternative energy source. They can be produced from renewable resources like corn, and they fit well with existing technology, such as gasoline- and diesel-powered cars. However, the use of biofuels made from crop plants also raises serious concerns:

- Crops grown for fuel may be more valuable than those grown for food.
  - If crop plants are used for fuel instead of food, food shortages will increase. (See the earlier section "Keeping up with human population growth" for more details.)
  - As food becomes more scarce, the cost of food will increase.
- Production of ethanol from corn requires a great deal of energy. The energy you get out of corn-based ethanol isn't much more than the energy it costs to make it.

Instead of crop plants, a better source for ethanol may be switchgrass, a perennial that requires less irrigation and fertilizer to grow.

Another creative idea for fueling our future is to use poop! Dairy farmers in Vermont are using their cow manure to produce the natural gas, *methane.* Methane can be used to generate electricity and to replace natural gas for heating. And, after all, no one wants to eat poop, so the competition with crop plants is avoided completely.

Discovering the Power of Plant Chemicals

Many manufacturers in resource-rich nations are promoting "natural" sweeteners and nutritional supplements. Natural plant extracts can be beneficial, but they can be dangerous, too.

Many of the world's most powerful medicines — and poisons — come from plants. In fact, sometimes the difference between a beneficial effect and a harmful one is just the amount of a particular chemical that a person is exposed to.

Finding a cure: Medicinal plants

People have used the bark and leaves of the willow tree to treat disease since ancient times. Chinese physicians used willow to relieve pain as early as 500 BC. Ancient Greeks, Egyptians, and Assyrians also used willow. For example, the Greek physician Hippocrates, who lived from 460 to 377 BC, suggested that women in labor should chew willow leaves for the relief of pain. People continued to use willow as a pain reliever and fever reducer throughout the world, and some people still use it today.

Although scientists and doctors discarded many ancient remedies as useless or even dangerous, the use of willow has stood the test of time. In fact, British chemists isolated the beneficial component of willow in 1828, a molecule called salicin. Scientists converted salicin to salicylic acid, and later to acetylsalicylic acid — in other words, aspirin!

Figure 20-4 shows a few of the world's most beneficial plants:

- White willow, whose scientific name is *Salix alba*, contains salicin, which relieves pain, and reduces fever and inflammation. Today, aspirin is made from acetylsalicylic acid which is synthesized in a lab.

- Foxglove, which botanists call *Digitalis purpurea*, produces digitalin, which is used to treat heart disease. Digitalin is used to make the medicines digitoxin or digoxin, which increase the force of heart contractions.

- Wormwood, or *Artemesia annua,* makes artemisinin, which has been used in Chinese medicine for many years under the name qinghaosu. It's become very valuable in recent years due to its antimalarial properties.

- Aloe is the transparent gel from inside the thick, succulent leaves of the *Aloe vera* plant. Aloe is used to treat wounds, burns, and other skin conditions.

- The happy tree, *Camptotheca acuminata*, is also called the tree of life and the cancer tree. This tree produces a compound called *camptothecin*, which has anticancer properties.

- Mormon tea, from the plant *Ephedra sinica*, has been used for many years in Chinese medicine under the name ma huang. The plant makes a chemical called ephedrine, which is a powerful stimulant and decongestant.
- The Pacific yew, *Taxus brevifolia*, makes the compound taxol in its bark. Taxol has anticancer properties and is particularly effective against breast and ovarian cancer.

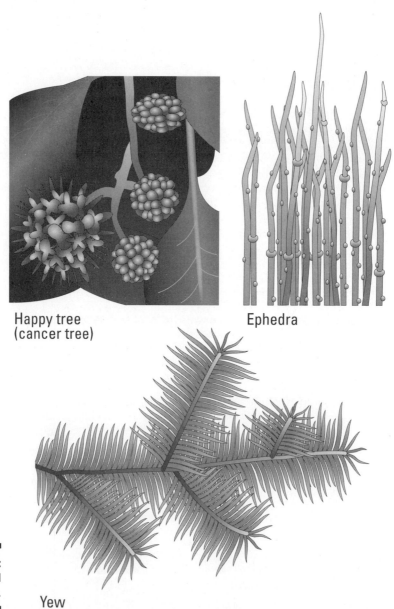

Figure 20-4: Medicinal plants.

Dangerous weapons: Poisonous plants

The famous Greek philosopher, Socrates, was sentenced to death for his beliefs and ordered to drink a cup of poison made from hemlock. The poisoning of Socrates is one of the most dramatic examples of power of poisonous plants, but many other more subtle examples surround us daily. Many common houseplants would be poisonous if you ate them, and animals, such as cows and sheep, can encounter poisonous plants while they graze.

Doctors have a saying, "The difference between medicine and a poison is the dose." Plant chemicals are powerful, and many of them could be poisons if used incorrectly. For example, the digitalin from foxglove could stop your heart if you ate too much. Of course, some plants are more dangerous than others.

Figure 20-5 includes a few of the world's most dangerous plants:

- Poison hemlock, or *Conium maculatum,* produces the poison coniine that is structurally similar to nicotine. Coniine affects the nervous system. In large doses, it leads to slowed heart rate, paralysis, and coma. Death is usually from respiratory failure.

- Many people grow the oleander plant, *Nerium oleander,* as an ornamental plant in their yard. And yet, oleander is ranked at the top of many lists of most poisonous plants in the world. Oleander makes two powerful toxins, oleanderin and neriine, which are most concentrated in the leaves. In fact, occasional poisonings result from children chewing the leaves of this plant. The toxin is similar to the digitalin found in foxglove. If eaten by people or other animals, the toxin causes vomiting, diarrhea, and abdominal pain, as well as an irregular heart rhythm and eventual coma and death.

- Locoweed, which is the name for some plants in the genus *Astragalus,* gets its name from the effect it has on cattle that eat it — *loco* comes from the Spanish word for crazy. Cattle that eat locoweed must have appeared crazy as they became aggressive, walked with a stiff and clumsy gait, held their heads low, and went blind so that they staggered around crashing into things before they died. Different species of *Astragalus* produce different toxins, but the toxin swainsonine causes the crazy symptoms in cattle when it affects cells of the central nervous system. In lower doses, the toxin can also lead to birth defects. On a good note, some of the chemicals from species of *Astragalus* have anti-cancer activity.

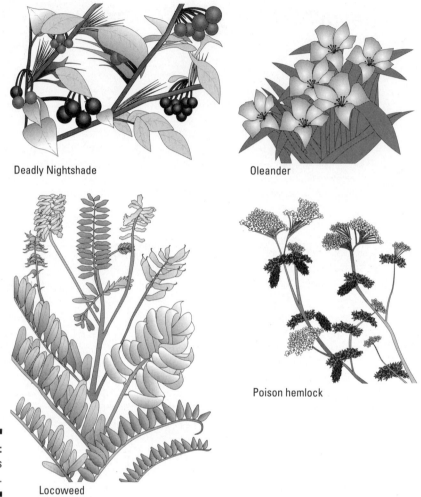

Figure 20-5: Poisonous plants.

- Deadly nightshade, or *Atropa belladonna,* produces shiny black berries that are sweet and juicy, occasionally tempting people to eat them. The berries and other parts of the plant contain two alkaloid toxins, atropine and scopolamine, which are deadly to humans and some animals. The poisons affect the nervous system, and if you eat enough of them, your involuntary muscles — including your heart — may become paralyzed. Before death arrives, your pupils dilate, and you'd experience sensitivity to light, blurred vision, headaches and confusion. When administered safely, however, the alkaloids from nightshade are beneficial — scopolamine is the main ingredient in a motion sickness patch, and atropine is used by opthalmologists to dilate their patients' eyes.

- Death camas are plants in the genus *Zigadenus*. They are delicate wildflowers that look like wild onion and often grow in areas where animals are grazing. Death camas produces alkaloid toxins that act on the central nervous system, leading to respiratory paralysis and death. Animals (especially sheep) that have eaten death camas are often found dead, lying near the plants they ate.

- Dumb cane, or *Dieffenbachia*, is considered one of the deadliest house plants and is also involved in livestock poisoning. The plants produce toxins called calcium oxalates in their sap. The toxin in the sap makes it very irritating to skin and mucus membranes. If the plant is eaten, the toxins cause burning and irritation of the mouth and tongue. The tongue may swell, leading to a loss of speech in people. If the tongue swells too much, it can block air flow, leading to death by suffocation. South American Indians used the irritating power of dumb cane sap when making poisons for the tips of their arrows.

- Jimson weed, or *Datura stramonium*, grows as a weed in pasturelands and soybean fields. All parts of the plant are poisonous, but the most common cause of jimson weed poisoning is when grazing animals eat the leaves. The plant produces powerful toxins, including the chemical atropine, which is the same toxin produced by deadly nightshade. In small doses, atropine slows down the heart and dilates the pupils. In larger doses, atropine may induce hallucinations and lead to cardiac and respiratory arrest.

- Coffee (*Coffea arabica*), tea (*Camellia sinensis*), cocoa (*Theobroma cacao*), and kola nuts (*Cola* sp.) all produce caffeine. You might wonder what they're doing on this list — after all, people drink caffeine all the time, and they have been for a long time all over the world in all type of societies. In the doses that people usually consume caffeine (say the equivalent of 2 cups of coffee), caffeine isn't toxic. But in high doses, caffeine can cause restlessness, insomnia, and tremors and can even lead to seizures and heart problems. So, although caffeine is familiar, people should still respect its power.

Having visions: Hallucinogenic plants

Many cultures and religions around the world have used hallucinogenic plants in their rituals.

Hallucinogens are chemicals that alter a person's state of mind so that they may see, hear, smell, taste, and touch things that aren't really there.

People have a long history of using hallucinogens, which are also called psychoactive drugs. Priests, medicine men, and other spiritual leaders have used hallucinogens because they think the altered state of mind helps them contact spiritual beings. Some churches use psychoactive drugs as part of their rituals. And some people take hallucinogenic drugs for recreational use.

Figure 20-6 show some plants that people use to alter their mental state:

Jimson weed

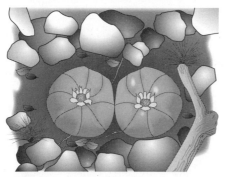
Peyote

Fly agaric

Marijuana

Magic mushrooms

Figure 20-6: Hallucinogenic plants.

✔ Peyote is a small, spineless cactus, with the scientific name *Lophophora williamsii*. Peyote contains the chemical mescaline. After the peyote is ingested, people often initially feel slight nausea. In addition to altered perceptions, mescaline causes increased heart rate and blood pressure.

- The fly agaric, or *Amanita muscaria,* isn't a plant at all; it's a fungus. It produces the hallucinogenic chemicals muscimol and ibotenic acid. These chemicals inhibit certain neurons in the brain and, in small doses, can produce feelings of euphoria, hallucinations, and drowsiness. Many people who eat these mushrooms fall into a deep sleep.

- Marijuana, or *Cannabis sativa,* has many names: People call it mary jane, pot, weed, and grass just to name a few. Marijuana's effects are caused by the chemical delta-9-tetrahydrocannabinol, or THC for short. Marijuana is usually smoked in order to take THC into the body. THC passes from the lungs to the blood and then to the brain and other organs. In the brain, THC binds to cells that have the right receptor for it. Receptors for THC are highest in the areas of the brain that influence pleasure, memory, thinking, concentration, sensory and time perception. Scientists have shown that the effects on memory and learning can last for weeks, so that people who smoke marijuana all the time may function at a lower intellectual level.

  Recently, the use of marijuana for medical purposes has caused renewed efforts to legalize its use. *Medical marijuana* is currently used primarily as way to relieve pain and side effects from anticancer drugs and from chronic disabling conditions, such as multiple sclerosis.

- Magic mushrooms, or mushrooms in the genus *Psilocybe,* produce the chemicals psilocybin and psilocin. After eating magic mushrooms, people often go through a period of intense yawning, followed by hallucinations and increased heart rate. The chemicals in these mushrooms have a very similar structure to the neurotransmitter serotonin, leading to stronger and longer serotonin signals in the brain.

- Black henbane, *Hyoscyamus niger,* is related to deadly nightshade and jimson weed. It produces similar alkaloid toxins, including atropine as well as hyosyamine. The difference between the toxicity of henbane and belladonna is mostly in the dose — henbane contains about ten times less the concentration of these chemicals. In the Middle Ages, Germans used to include henbane, called bilsen, as an ingredient in beer. Bilsen became pilsen over time, but is no longer included in pilsen beer because of its toxicity. People who thought they were witches may have spread a salve containing belladonna and henbane on their skin, causing them to have visions of flying in the air.

- Mandrake, or *Mandragora officinarum,* is famous because of its weird looking roots that people thought looked like, well, people. In the Middle Ages, many people believed in the Doctrine of Signatures, which said that the way a plant looked gave a clue to its medicinal uses. So, mandrake, which had roots that looked like the human body, was supposed to cure just about everything. People also believed that when you picked mandrake, the screams of the roots could knock a person out. Legends about mandrake persisted for centuries and even made it into the Harry

Potter books. Like its relatives, belladonna and black henbane, mandrake produces alkaloids like hyoscyamine, atropine, and scopolamine. Like black henbane, people who believed they were witches used mandrake in their potions, probably resulting in hallucinations.

- Kwashi, which scientists call *Pancratium trianthum,* is used by the Bushmen in Botswana. To induce visual hallucinations, the bushmen rub the bulb of the plant over cuts in their scalp. The active chemical in kwashi isn't known, but some relatives of kwashi produce alkaloids.

Part VI
The Part of Tens

In this part . . .

On the surface, plants may seem like they're not doing much but standing around looking pretty. But when you take a closer look, you'll discover there's more to plants than meets the eye! Plants trick animals into "mating" with them, trap and digest animals, mimic the appearance (and smell!) of rotting meat, and do many other amazing things. Studying plants can be surprising and fun, but it can also be hard. In this part, I share both the lighter side of plants and also ten tips for getting a handle on the subject and getting a good grade.

Chapter 21
Ten Weirdest Plants

In This Chapter

▶ Checking out the tricks plants use to control insect behavior

▶ Discovering that there's more to plant life than photosynthesis

▶ Marveling at the strategies plants have for protecting themselves

Plants just sit there, so they're kinda boring, right? Not so! This chapter presents ten ambassadors from the plant kingdom to prove it to you.

Plants That Eat Insects: Cobra Lily

Plants that eat insects have many clever tricks to lure and trap their next meal. The cobra lily, *Darlingtonia californica*, is a master of silent death. The leaves of the cobra lily are folded into a tube that bends over, forming a shape like the head of an upraised cobra. It even has two little flags of leaf that stick out to the sides at the bottom of the head, looking a little like fangs.

But while the whole cobra-like appearance is cool, that's not what tricks the insects. They're lured by the sweet smell that the cobra lily releases. The insects investigate and find an opening in the base of the curved tube, near what would be the cobra's mouth. They enter the leaf and are disappointed to find nothing to eat, so they decide to leave. The insects look for the light to tell them where the exit is, and they see little windows of light at the top of the plant. They fly toward the light, only to find that these are false windows, mere thin areas in the leaf that let the light through but aren't actually open to the outside. The confused insects beat their heads against the windows, trying to fly out. They grow tired and fall back to the slippery sides of the leafy tube. If they decide to try to climb out, they're confronted with sharp hairs that point downward at them from the sides of the leaf and a slippery substance on the sides of the tube. The insects may try climbing for a while or switch back to flying, until, exhausted and hopeless, they fall to the bottom of the leaf. Then, at night, the cobra lily raises the level of water — water plus digestive enzymes and bacteria — breaking down its insect meal. The cobra lily silently absorbs the minerals it needs from the corpses of its prey, patiently waiting for a new day to bring it more victims.

Plants That Stink: The Corpse Flower

The corpse flower, scientifically known as *Amorphophallus titanum,* is pollinated by flies. And what do flies like? Nasty things, like rotting meat. When the corpse flower blooms, it grows a massive spear-like inflorescence that grows up to around six feet in height and is a dull, red, meat-like color.

And because the visual isn't enough to get the job done, during the peak of its bloom the flower puts out an enormous stench designed to convince the flies that the dull-red stinky inflorescence is actually the real thing, and they should come check it out. And when I say enormous, I mean mind-boggling amount of stench. I've heard stories of the plants being moved outside of the greenhouses where they live when they're at the peak of their powers because the people who work in the greenhouses can't take the stink. I've also seen pictures of people standing next to *Amorphophallus* in bloom — wearing gas masks! And yet, when one of these plants blooms, which only happens every few years, it often makes the newspaper and people — literally, thousands of people! — come from miles around to see it. After all, it's not every day you can see the largest, smelliest plant in the world.

Plants That Move: Galloping Moss

Many plants move — Venus flytraps snap shut on their pray, sensitive plants fold up their leaves, vine tendrils swivel to find something to hold on to. But as a representative of plants that move, I've chosen a plant that actually travels from one place to another. A cute, spiky little green moss, named *Grimmia ovalis,* actually slides across the surface of the ground, traveling slowly downhill over the surface of lichen-encrusted rocks in the Arctic. The movement is very slow, and scientists think it may be propelled by the freeze-thaw cycle on the rocks (so maybe I'm cheating a little), but the mosses leave behind a clear trail that shows where they've been. So, when you look at the rocks, it looks like a moss race track, with clumps of moss trailing streaks of cleared rock behind them. Which is totally cool. Other plants that are growing on the moss clump get a free ride. I can imagine these mosses speeded up a little bit and starring in a children's book. As Dr. Seuss might say, "From here to there and there to here, galloping moss is everywhere."

Plants That Mimic Rocks: Stone Plants

Plants have many strategies to avoid predation, like thorns, spines, and nasty secretions. One plant that's a master of the subtle defense is the stone plant, genus *Lithops (litho-* means stone). Stone plants are rock-colored — dull greys and tans — and have two thick succulent leaves that are broad and flat at the top. The plants live in the desert, literally buried up to their necks — okay, the top of their leaves — in the desert soil. Only the flattened tops of the rocks are visible on the soil where they look pretty much like rocks. Any wandering animals that might be looking for juicy plants in order to get some water will just walk on by.

Of course, being buried in the soil presents some problems, such as getting enough light for photosynthesis. But the stone plants solve this problem by having thin spots in the tops of their fleshy leaves, little windows in effect, that let light into the interior of the leaf. Photosynthesis actually happens within the leaves below the soil. Stone plants have other names, too, including pebble plants and living stones, that illustrate just how successful their rock mimicry is.

Plants That Come Back From the Dead: Resurrection Plant

Well, the resurrection plant actually comes back from the "looks dead." But it's pretty amazing just the same. The plant, which scientists call *Selaginella lepidophylla,* lives in the deserts of North and South America. When it rains, the plant grows rapidly, producing a rosette of green branches. When it's not raining, the plant starts to curl up, and its outer branches dry out. The plant continues to roll up until it looks like a ball of dead, dried up branches. If you found this plant in this state, you'd think it was dead and fried by the sun. But inside the center of the ball, the plant is conserving its last bit of water. The branches are dormant, but not dead, and can stay in this dormant, dried up state for years. When the rains return, however, it turns green again and begins to unfold and grow. This transformation is so dramatic that gift shops in the American Southwest sell the dried up resurrection plants, sometimes called the Rose of Jericho, as souvenirs.

Plants That Just Look Strange: Welwitschia

There's nothing quite like a *Welwitschia* plant. The plants grow only two leaves from a single stem for their entire lives, which usually last about 500 years! And during that time, the two leaves keep growing, and growing, and growing.

The plant is low to the ground — the stem bases get about two feet high or so — and the big, strappy leaves keep growing outward from the base, getting longer and longer, like wide green ribbons flowing away from the central stem. The leaves get torn and split over time, and they curl and twist in interesting arrangements around the stem. The plant ends up looking like some large, strange, alien creature that might start walking across the desert toward you.

Welwitschia is a gymnosperm, so it makes cones when it reproduces. Each plant is either male or female, but not both. When the plants are reproducing, they form a ring of cones in a ring around the stem base. These stick up looking a little like the antennae or stalked eyes, really completing the alien creature look the plant seems to be going for.

Plants That Aren't Green: Indian Pipe

Everyone knows that plants are green. Even if they have other colors, too, somewhere they're green. Not so with Indian pipe. These small, ghostly looking plants are pure white. The plants have slender stems that grow to reach between five and ten inches in height, and they have a single, delicate, bell-shaped flower that nods downward on the stem. The plants are so white, they almost look like wax; some people even call them *corpse plants* because of this white, waxy appearance. Other people mistake them for fungi. But, no, they're plants.

You might wonder how the Indian pipe survives; after all, no green means no chlorophyll, means no photosynthesis, which means no food. The plants have very cleverly solved this minor difficulty by tapping into a fungus that lives in the soil. The plant's roots connect to the fungal cells that are growing as threads throughout the soil.

And this is not just any fungus. It's a type of fungus called *mycorrhizae* that has a special relationship with a nearby tree. So, the Indian pipe taps into the fungus, which is tapped into a tree. The fungus and tree have a working relationship — the tree gives the fungus sugar from photosynthesis, and the fungus helps the tree get water and minerals. The Indian pipe doesn't help anyone do anything; it just taps into the fungus and helps itself to all the goodies that the tree and fungus are sharing.

Plants That Have Sex with Insects: Australian Tongue Orchid

I just love orchids, don't you? Their flowers have such interesting shapes and colors, and they make me think of exotic locales. But no matter how much you or I love orchids, we don't love orchids the way orchid dupe wasps do. In particular, the wasps love a species of orchid called the Australian tongue orchid, scientific genus *Cryptostylis*. The orchids lure the male wasps with their scent and appearance, which is styled to mimic a female wasp. The orchids that have the best copycats are most likely to get pollinated by the visiting wasps, so over time, the orchids have gotten very, very good at mimicking a female wasp. The male wasp smells the alluring scent of female wasp, looks around and sees (with its ability to see ultraviolet light) patterns on the orchid flower that look just like the patterns on a female lovely of its own species. The wasp concludes that the orchid is the source of the enticing, come-hither smells, so it buzzes on over.

Until recently, scientists thought that the wasps just attempted to mate with the orchid, but then would get discouraged and leave (carrying their load of pollen, which was the goal of the plant). But one scientist who was studying the wasp-orchid relationship thought that the mating looked pretty close to the real thing. So, after wasps were finished visiting flowers, the scientist and some of her colleagues rushed the orchid flowers to the lab to examine just how far things really went. When they looked into the flowers with a microscope, they saw swimming wasp sperm! I guess the only question remaining is how long the relationship will last.

Plants That Really Know How to Grow: Queen Victoria Water Lily

The leaves of the Queen Victoria water lily *(Victoria amazonica)* can grow to over 8 feet in diameter! And the leaves grow fast — at one horticultural pond, the leaves increased from covering 5.9 square feet of water surface to 35 square feet in just a two-week period. The leaves float on the surface of the water with the help of the network of veins that form a web over the back of the leaf. Pockets of air form between the veins, acting like little balloons to float the leaf. The back of the leaf is also covered with vicious spines that help protect the leaves from predators that live in the water.

When the lily blooms, it makes a lovely white flower about the size of a soccer ball. The flower opens at night as a female, heating up to send its pineapple-like scent out to attract scarab beetles to enter the flower. In the morning, the flower closes, trapping any visiting beetles inside. The next night, the flower re-opens, but as a male flower that is now pink. As the beetles leave the flower, the anthers of the male flower release pollen, dusting the beetles so that they'll carry pollen to the next flower.

As beautiful as the flowers are, it's the leaves of the Victoria lily that people like to see. If you search for this plant on the Internet, you'll find pictures of the enormous leaves with children sitting on them. The leaves are so large that if you distribute the weight of a child across the surface, they can easily float there.

Plants That Climb: Banyan Tree

The rainforest is lush and green, so crowded with plants, in fact, that competition for light can be fierce. Plants grow on top of other plants in order to make their way into the tree canopy where the light is. Some plants spend their whole lives in the canopy as epiphytes, never putting roots to soil. Other plants start out in the soil and then grow upward as vines until they reach the canopy. But the banyan tree has an unusual strategy that combines a bit of both.

Banyan trees start off up in the canopy, germinating on branches of other trees from seeds that have been dispersed by birds. At first, the banyan seedlings live as epiphytes, using their roots to collect water and nutrients up in the canopy. The roots of the banyans, which are prop roots, continue to grow slowly downward, twisting around the host tree as they go. Once the roots reach the soil, everything changes. The banyan trees grow faster, sending out a network of roots that fuse together and surround the host tree. The banyan outcompetes the host tree for water and nutrients. The leaves of the banyan tree spread out, cutting off light from the host tree. The host tree dies, strangled to death by the banyan, which is why the banyan is also called the strangler fig. The host tree gradually decomposes, leaving just a hollow core on the inside of the banyan. Banyan trees can get very large — one tree in Burma grew to cover 4 acres and made 1,000 prop roots.

Chapter 22
Ten Tips for Improving Your Grade in Botany

In This Chapter
▶ Figuring out the best strategies for lecture and lab
▶ Perfecting your study habits
▶ Challenging yourself with different kinds of questions

Botany is a science, and like all sciences, it requires you to memorize a lot of specialized terminology and detailed processes. To do well, you need to find the best study strategies that work for you. In this chapter, I present ten tips to ensure that you're using your time effectively and getting the best grade possible.

Listen Actively to Lecture

You can easily lose your focus during a lecture, especially if the instructor doesn't ask many questions. And if lots of information is delivered quickly, it can be overwhelming. But lecture is your chance to hear what's important to the instructor and how she makes sense of the material. So, here are a few tips to help you get the most out of the lecture part of your class:

- **Preread the chapter before coming to lecture.** If you read the chapter and take notes or even jot questions in the margin of your text, then you'll be more prepared to follow along and ask questions during class.

- **Write notes in your own words.** This advice can be challenging, especially today when many instructors present information using PowerPoint and may even give you outlines of everything that's on their slides. In this kind of situation, you can easily to kick back and relax — maybe even to get too relaxed and tune out. To keep yourself in the moment, listen to what your instructor is saying and write some notes in your own words.

Just because you understand what your instructor is saying doesn't mean that you'll do well on the test. Understanding in lecture just proves you have the capacity to comprehend the material when someone explains it to you. It doesn't mean you know it! A better test of whether you've really learned anything is whether you can explain it to someone else (or yourself when you're studying).

- **Take notes on interesting stories and anecdotes.** Instructors often tell stories and give examples to show the relevance of the information they're presenting. If the instructor tells a good one that helps you grasp the concept he's talking about, jot down a few notes about it. When you're studying later, these side notes may help you recall the topic.

- **Sit in the best place for you in lecture.** Usually, the front is best because it's too easy to get distracted and tune out in the back. However, if you're someone who gets sleepy and might need to move around a little to wake up, then try an aisle seat. If you get sleepy, you can get up and take a short walk to the restroom rather than fall asleep.

- **Ask questions when you don't understand.** If you're following the lecture but something doesn't make sense to you, then ask about it. Chances are if you don't get it, someone else doesn't either.

Use Your Lab Time Effectively

Lab is an opportunity for you to experience the material in another way, which can really help you learn. But lab also has its own challenges, especially if you have to prepare to take a lab exam called a *lab practical*.

When you're in lab, be sure to

- **Make clear drawings that are useful to you when you study.** Clear, useful sketches are particularly important if you have to take a lab practical where you're shown samples from your previous lab work and asked to answer questions about them. Often, the only thing you'll have to study from is your own lab notes and drawings. You don't have to be a gifted artist to make useful lab drawings — you just need to draw fairly clearly, use lots of labels, and make notes to yourself to help you remember. For example, if you're drawing pine pollen and you think it looks like Mickey Mouse, you could draw pollen, label your drawing, and jot a note off to the side of your page that says, "Pine pollen looks like Mickey Mouse." Then when you're studying for your lab practical, you can study your own tips and clues. So, when you take the practical and look at a slide of pine pollen, even if you don't remember your drawing very well, you might remember your tip to yourself and successfully identify the pollen because it looks like Mickey Mouse.

- **Get as much help from the teaching assistant or professor as you can.** Depending on where you go to school, your professor may not teach the lab section of your course. If you attend a big school, a graduate teaching assistant often runs the lab. Teaching assistants are often easier to approach than professors, and they know a lot about what you're learning. So, ask them for help with lab stuff and with lecture material you don't understand. You can ask them these questions when you have time in lab or during their office hours.

- **Don't leave lab early unless you've done everything fully and completely, including answering all the questions.** Sometimes you may be tempted to cut things short and head out early, but if you do, you're wasting precious learning time. Lab gives you the opportunity to look at examples of what you're learning, touch them, and discuss your ideas with your lab partners and your professor or TA.

- **Choose your lab partners wisely.** If you're easily distracted into chatting about nonbotanical subjects, don't partner up with people who'll encourage that behavior. Try to partner up with people who do the lab work thoroughly and cooperatively and who'll stay focused on the subject for the lab.

Plan Your Study Time

The general rule for college science classes is that you should spend two hours outside of class studying for every one hour you're in class. So, the number one thing you should do every week is to make sure that you reserve enough study time. You'll get the best bang for your study time if you're smart about how you schedule it:

- **Plan for studying like you do for other important events.** Make studying a priority and figure out how to fit it into your life.

- **Schedule enough time each week.** Science classes typically require more study time than other types of classes.

- **Schedule short blocks of time each day as well as longer blocks a few times a week.** Studies show that you get the most out of studying your lecture notes the sooner you review them after lecture. So, plan some study time every day so that you can do a review before you forget what you heard. When you sleep at night, your brain edits your experiences for the day and stores only some things in long-term memory. By studying soon after lecture, you give yourself the chance to clarify your notes before you forget material.

- **Find the best study strategies for you.** Some people learn best from pictures, some from sounds, and some from moving around. You may have one way that works best for you, but chances are you can learn from all these approaches. So, when you're studying, mix it up a bit. Draw pictures, talk out loud, act out processes using the objects on your desk, or get up and move around. Also, relate the information you're learning to your own life or things you've heard in the news. The more neurons you can engage while you're studying, the better your understanding is likely to be.

Be Active, Not Passive

If you understand what you're hearing in class and reading in your text, that's great! Understanding is one of the first steps to learning the material. But it's only a first step. Unless you have a photographic memory, you're going to have to do some work to make the knowledge your own:

- **Put things in your own words.** Memorizing a definition from the book or an explanation from instructor will only get you so far. You can say the same thing many different ways, and your instructor is likely to use different ways of phrasing things in lecture and on exams.

- **Alternate between studying the information and testing yourself.** After you study a topic, try to recreate it for yourself. Practice drawing diagrams on blank pieces of paper, looking at your notes only when necessary, or explain something to your study partners or your mom. The key is that you're alternating between information storage and information retrieval. Studies on learning have shown that retrieval practice, such as writing out explanations, helps improve recall.

- **Practice applying the information to new situations.** Once you think you've got it, try solving every problem you can get your hands on. Solve the harder problems at the end of the chapter in the text, review any problems given in class, and do any homework that was assigned.

Make Up Tricks to Jog Your Memory

Like most science classes, botany is full of new terms and processes you'll need to learn. There's really just no way around it. One way you can help yourself is to make up words or sayings to help trigger your memory of terms and events. For example, way, way, back in the fourth grade, I learned HOMES to represent the Great Lakes of the United States. To this day, I still know HOMES represents Huron, Ontario, Michigan, Erie, and Superior. You can make up your own memory tricks, get them from friends, or look online.

Prepare for Different Kinds of Questions

If you're taking a course in botany, then you're probably not an expert in the subject. To get from beginner to expert, you need to do a lot of learning on different levels. And, you need to be prepared to be tested at various levels of understanding. On an exam, you may be asked to

- **Recall facts.** You have to learn many terms for different structures and processes in molecular and cellular biology. Before you can move on to higher levels or even understand your instructor, you need to learn the language of the subject. You'll be tested at this level by questions that ask *what* things are. Questions that test factual knowledge may ask you to "name," "recall," "identify," or "label." To prepare for recall questions, you can make flash cards and test yourself, practice labeling drawings, or make lists of the characteristics of plant groups.

- **Demonstrate understanding.** You may be asked to describe a process or explain something in your own words. You can prepare for these types of questions by describing a process out loud, drawing it on a blank piece of paper, or think of real-life examples that illustrate a process.

- **Apply information to solve new problems.** You may be asked to predict the outcome of something or identify the problem in a given situation. When you're studying and you think you understand a process, ask yourself, "What would happen if I changed this component of the system?" If your instructor used graphs to explain certain concepts, be sure to study the graphs and redraw some of them so that you understand how the graph illustrates the concept.

Remember the Supporting Material

Textbooks come with lots of good and useful supporting materials these days — CDs, websites, practice tests, student study guides, and more.

The supporting materials that came with your text were probably written by someone other than the textbook author or your instructor and may not be at the same level as your course. If a practice test is heavy on recall, but your instructor tests at a higher level of understanding and application, the practice test may give you a false sense of confidence before an exam. Definitely take advantage of whatever materials come with your text, but you need to evaluate them carefully first! If your instructor recommends a specific website, be sure to check it out.

Test Yourself Often

The big test day will come, and you'll get a permanent grade. Before you get to that crucial moment, make sure that you've put yourself through your paces in a less stressful situation. Test yourself during study sessions and get together with other students in the class to test each other. Make sure that when you test yourself, you're practicing different types of questions. (See the earlier section, "Prepare for Different Kinds of Questions.")

If your instructor releases old exams, use them as practice tests after you've studied the material and think you're ready. Don't just read the practice test and the answers — it's too easy to think "Oh, yeah, I get that." Sit down and take the test like it's the actual test and see how you do.

Use Your First Test as a Diagnostic Tool

You won't really know how your instructor tests until you take your first test, unless you've had him in the past (or know someone that has). Some instructors include recall questions in their tests, while others just write problems for you to solve.

Your first test is a very valuable piece of information. When you get your graded test, go over it and look for the types of questions your instructor asked. What kinds of questions were you asked? Pay particular attention to any questions you missed and ask yourself why you missed it. What type of question was it — recall, concept, application? Where was the information? Does your instructor test only off notes? Or does he test on concepts in the book that he didn't cover in class? Is information learned in lab covered on exams?

The more you know about what to expect, the more you can tailor your study sessions to fit your instructor. Study smarter, not longer!

Get Help Sooner Rather Than Later

If you're doing your best to actively follow along in lecture and you're not getting it, go for help right away even if it's the first day of class! Your instructor and teaching assistants have office hours and e-mail; don't be afraid to use them! Many schools have tutoring services as well.

Index

• A •

abiotic factor, 291
abscisic acid, 166, 170, 177
acacia tree, 302
accessory fruit, 95
accessory pigment, 119, 124
accessory tissue, 95
acetyl-coA, 134, 139–143
achene, 94–96
acid, 18–19, 253, 309
acid-growth theory, 169
actin, 37
actinomorphic flower, 282
active learning, 352
active site, 104
active transport, 150–151, 160–162
adaptation
  access to water, 197–198
  aquatic-to-land transition, 244
  described, 219, 223–224
  examples of, 223–230
  mutations, 218
  natural selection, 218–221
  plant life cycles, 197–198
  seedless vascular plants, 253–254
adaptive radiation, 219
adenine, 24, 26
adenosine diphosphate (ADP)
  defined, 105
  energy transport and storage, 105–107
  glycolysis, 136
  Krebs cycle, 142
  photosynthesis, 123
adenosine triphosphate (ATP)
  cellular respiration processes, 133–138
  chemiosmosis, 121–122, 143–146
  described, 39, 105
  energy pyramid, 296–297
  energy transport and storage, 105–107, 117
  glycolysis, 134–138
  heat generation, 144
  illustrated, 106
  Krebs cycle, 139–143
  light independent reactions, 125–129
  photophosphorylation, 122–125
adhesion, 155
ADP. *See* adenosine diphosphate
adventitious root, 68, 73
aerenchyma, 48
aerial root, 73
agamospermy, 183, 184
agave, 224, 225
age, of tree, 65, 66
aggregate, 94–95
agricultural history, 323–326
Agrobacterium, 314–317
air chamber, 247
a-ketoglutarate, 142
alder tree, 303
aleurone layer, 178
aleuroplast, 39
algae
  absorption spectra, 115
  evolution, 244
  photosynthesis, 122
  plant kingdom, 239–240
  symbiosis, 303
algal bloom, 310
alkaloid, 284
allele, 204–214
allelopathy, 304
allergen, 322
aloe, 79, 333
alpha-amylase, 178
alternate leaf, 77, 79
alternation of generations, 196–198
alternative energy, 113, 119
amensalism, 304
amino acid, 22
ammonia, 300, 301
ammonification, 301
amyloplast, 39, 172
anabolism, 102, 106
anaphase
  meiosis, 193, 194
  mitosis, 189
ancient civilization, 323–324
androecium, 89
angiosperm. *See also* flowering plant
  described, 275, 278–279
  diversity, 281–286
  evolution, 275–278
  versus gymnosperms, 82–83, 266–267
  life cycle, 279–281
  pollination strategies, 278, 286–288
animal
  cell components, 34
  cell cycle, 194
  genetic modification, 322
  pollination ecology, 286–288
  seed dispersal, 97, 98, 270
  symbiosis, 302
anisotomous, 256
annual plant, 63
annulus, 263
ant, 302
antenna complex, 119–120
anther, 88, 279
antheridia, 247–248, 252, 259, 263
antheridiophore, 248
Anthocerophyta (phylum), 249
apical cell, 247
apical dominance, 167
apical meristem
  described, 11, 45–46
  location of, 58
  root components, 71
  root growth, 69, 164
  stem growth, 59–65
apoptosis, 186
apple, 95
aquatic plant, 229–230, 244, 347
Arabidopsis thaliana, 177, 206, 218
arable land, 326
archaea, 30, 233

archegoniophore, 248, 249
archegonium
    liverwort life cycle, 247–248
    lycophytes, 258
    moss life cycle, 252
    pterophytes, 263, 264
artemisinin, 333
artificial selection, 219
asexual reproduction, 182–185, 248
astrosclereid, 48
atom, 14–18, 103–104, 134
ATP. *See* adenosine triphosphate
ATP synthase, 121, 124, 125, 145
atropine, 336, 337
Australian tongue orchid, 346–347
autotroph, 102, 103, 293
auxin, 165–170, 172
axil, 59
axillary bud, 46, 59, 78

• B •

Bacillus thuringiensis (*Bt*), 318–319
bacteria
    described, 29, 233
    epidermis functions, 51
    nitrogen cycle, 300–301
    photosynthesis, 118
    phylogenetic trees, 235
    symbiosis, 303
    wooden cutting boards, 68
bacteriochlorophyll, 115
banner, 283
banyan tree, 348
bark, 64, 72
barrel cactus, 224, 225
basal angiosperm, 277, 282
basal rosette, 77, 79
base, 18–19
bat, 288
bean, 83–84, 170
bee, 90, 287
beer, 339
beetle, 90, 288
belladonna, 340
berry, 94–95, 270
beta-carotene, 318
binomial nomenclature, 240–242
bioaccumulation, 310
biodiversity, 310

biofortification, 318
biofuel, 330–332
biogeochemical cycle, 298
biological evolution, 217
biological nitrogen fixation, 300
biomass, 304–305
biome, 305–310
biosphere, 291
biosynthesis, 126
biotic factor, 291–292
bird, 288
black henbane, 339, 340
black pepper, 276, 277
blade, 75
bleach, 329
blight, 303
blue light, 177
Bohr's model, 15
boreal forest, 306
Borlaug, Norman (scientist), 326
botany, 1, 7
bract, 79, 80
branch root, 159
branch, tree, 256
Brassica oleracea, 78
brassinosteroid, 28, 166, 171
breathing root, 73
bristlecone pine, 66
Bromeliad plant, 226–227, 302
Brownian motion, 149
Brussels sprouts, 78
Bryophyta (phylum), 246, 250–253
bryophyte, 244–246. *See also specific types*
Bt (*Bacillus thuringiensis*), 318–319
bud, 58–59, 87–88
bulb, 67, 79–80
bulbil, 258
bulk flow, 162
bundle scar, 59
buttercup, 283
butterfly, 288, 318, 319
buttress root, 73, 226

• C •

C_3 plant, 129
C_4 plant, 127–129
cabbage, 78, 170
cactus, 127–128, 225
caffeine, 337
calcium oxalate, 337

callus, 313
calyptra, 249, 250, 252
calyx, 88, 89
CAM (Crassulacean acid metabolism) plant, 127–128
cancer, 271, 333, 334
capillary action, 155
capillary tube, 155
capitulum, 92, 284
capsule, 94, 96
carbohydrate
    described, 20–21
    light independent reactions, 125–129
    photosynthesis process, 116, 118, 125–129, 131
    stored energy, 27
carbon
    atomic mass/number, 16
    cycle, 298–300
    elemental structure, 15
    fixation, 125–126, 127, 298
    Krebs cycle, 139–143
    light independent reactions, 125–129
    lipids, 26
    sinks, 299, 300, 326
    sources, 299
carbon dioxide
    carbon cycle, 298–299
    cellular respiration overview, 132, 133
    epidermis functions, 51
    fossil fuel burning, 300
    Krebs cycle, 139–143
    photosynthesis, 116, 117, 118, 125–129, 131–132
    pollution, 27
    transpiration, 157
carbon-hydrogen bond, 27
Carboniferous period, 257, 295
carnivore, 293
carnivorous plant, 228–229, 343
carotenoid, 39, 115
carpel, 88, 278
carrion flower, 288
caryopsis, 85, 94, 96
Casparian strip, 71, 160
catabolism, 102, 104–106
cation, 159–160
catkin, 91, 92
cavitation, 158–159
cell cycle, 185–186, 191, 194
cell division. *See* mitosis

Index

cell elongation
  auxins, 168–169
  described, 163–164
  tropisms, 172–173
cell, plant
  basic components, 29–33, 35
  described, 29
  differentiation, 163–164
  formation of tissues, 43
  growth and development, 163–165
  importance of, 13
  macromolecules, 19–28
  seeds versus spores, 82
cell plate, 189
cell wall
  collenchyma, 49
  described, 23, 40–42
  illustrated, 29
  layers of, 41
  parenchyma, 47
  sclerenchyma, 49
  vascular bundles, 60
cellular respiration
  basic process, 39
  chemiosmosis, 143–146
  defined, 10, 39
  function of, 131
  glycolysis, 133–138
  heat generation, 144
  illustrated, 134
  versus photosynthesis, 131–132
  summary reaction, 132
cellulose, 21, 41
centromere, 186
channel protein, 31
charophyte, 239–240, 244
checkpoint, 186
chemical. *See also specific chemicals*
  bond, 17–21, 103–104
  elements, 14–17
chemical reaction
  cellular respiration, 132
  enzyme function, 104
  glycolysis, 133–138
  metabolic pathways, 103–104
  photosynthesis, 112, 116, 119–120
  redox reactions, 107
chemiosmosis, 121, 143–146
chestnut tree, 303
chlorenchyma, 48
chlorophyll
  absorption spectra, 114
  defined, 37, 114

function of, 114
light reactions, 119–125
photophosphorylation, 122–125
solar energy uses, 119
structure of, 115
types of, 115
chloroplast, 29, 37–39, 249
chromoplast, 39
chromosome
  described, 25, 187, 199, 200, 204
  evolution, 223
  incomplete dominance, 212–214
  interphase, 186
  meiosis, 190–194, 211–212
  mitosis, 187–189
  multiple-gene inheritance, 207–212
  single-gene inheritance, 199–207
cilia, 37
circadian rhythm, 176–177
citrate, 142
citric acid cycle, 134, 139–143
clade, 236
classical systematics, 237
classification, plant, 236, 240–242, 260
climbing plant, 173, 348
clone, 170, 182
club moss, 255–259
cobra lily, 343
coconut, 95
coenzyme A, 140
coevolution, 287
cohesion, 154–156
cold climate, 177–178
coleoptile, 85, 166–167
coleorrhiza, 85
collagen, 41
collenchyma, 48–49, 61
color, flower, 90, 288, 289
commensalism, 304
community, 292, 305–310
companion cell, 54–55
competition, 304
complex carbohydrate, 20
complex tissue, 47
compound leaf, 77–78
compound ovary, 88
concentration gradient, 149–150
condensation, 187
cone, 79, 87
conifer, 60, 72, 267, 270

coniine, 336
conservation biology, 310
consumer, 293, 298, 300
control group, 167
convergent evolution, 224
copper, 16
cork
  cells, 52, 64
  defined, 46
  forests, 65
  periderm, 52
cork cambium, 46, 64, 72
cork oak, 65
corm, 67
corn
  agamospermy, 183
  ethanol, 331–332
  genetic engineering, 318–319
  germination, 85–86
  importance of, 183
  reproduction, 93
corolla, 88, 89
corpse flower, 344
cortex, 61, 70
cotton, 319, 330
cotyledon, 83–86
covalent bond
  atomic structure, 15, 17, 18
  lipids, 26
  nucleic acids, 24
Crassulacean acid metabolism (CAM) plant, 127–128
cristae, 40
crocus, 67
crossing-over process, 192, 207–212
cross-pollination, 287
cryptochrome, 177
curvature response, 173
cuticle, 50, 75
cutin, 50
cutting board, 68
cyanobacteria, 115, 118, 122
cycad, 267, 268–269
cyclic pathway, 103
cyclic photophosphorylation, 122, 124–125
cytochrome, 123
cytokinesis, 187, 194
cytokinin, 165, 169–171
cytoplasm, 29, 31, 154
cytoplasmic streaming, 37
cytosine, 24, 26
cytoskeletal protein, 22–23
cytoskeleton, 36–37

• D •

dark environment, 143, 176, 178
Darwin, Charles (scientist), 219–221
daughter cell, 189
day neutral plant, 175
deadly nightshade, 284, 335, 336
death cama, 336
decarboxylation, 140
deciduous plant, 59
decomposer, 293, 297, 298, 301
deforestation, 300, 326
dehiscent fruit, 96
de-inking, 329
denitrification, 301
deoxyribose, 24
dermal tissue system, 44, 50–52
descent with modification, 219
desert
  adaptations, 224–225
  biomes, 305–306
  cavitation, 158
  photosynthesis, 127–128
  transpiration, 157
desert yam, 224, 225
determinate growth, 45
detritus, 293
Devonian period, 266
dichotomous key, 241–242
dicot
  angiosperm life cycle, 280
  described, 60, 282–284
  flower structure, 88–89
  germination, 84–85
  growth of, 63–65
  leaf tissues, 76
  root systems, 70–72
  secondary growth, 72
  seed structure, 83–85
  stems, 60–63
dictyosome, 29, 35
differentiation, 163–164
diffusion, 149–150
digitalin, 333
dihybrid cross, 207–212
dihydroxyacetone phosphate, 138
dimorphic plant, 261
dioecious plant, 248
diploid, 83, 190, 200
disk flower, 92, 284
dissaccharide, 20
disturbance, 307–310
divergent evolution, 223
division, 238
DNA
  angiosperm evolution, 277
  components of, 23–26
  described, 23, 31–32
  genetic engineering, 314–317
  incomplete dominance, 212–214
  interphase, 186
  meiosis, 191–194
  mitosis, 187–189
  multiple-gene inheritance, 207–212
  mutations, 218
  natural selection, 220
  plant kingdom, 239–240
  shared characteristics, 231–232
  single-gene inheritance, 199–207
DNA ligase, 315
domain, 29, 232–235
dormancy, 170, 177
double compound leaf, 78
double fertilization, 279
drawing, 350
drip tip, 226
drupe, 94–95
dry fruit, 93, 96–97
dry milling, 332
duckweed, 230
dumb cane, 337
dwarf plant, 170

• E •

ECM (extracellular matrix), 41
ecological efficiency, 297
ecological niche, 292
ecologist, 292
ecology, 292
ecoplast, 39
ecosystem
  biomes, 304–310
  carbon cycle, 298–300
  described, 291–292
  energy flow in, 293–297
  nitrogen cycle, 300–301
  role of species, 292–293
  symbiosis, 302–304
ectomycorrhizae, 74
egg
  angiosperms, 279
  defined, 9
  liverwort life cycle, 248
  lycophytes, 258, 259
  moss life cycle, 252
  pines, 272
  plants versus animals, 195
  pterophytes, 264
  role of, 184
  trait predictions, 205–207, 209–211
elaioplast, 39
elater, 249
electromagnetic radiation, 113–114
electron
  carriers, 107–108
  cellular respiration overview, 133
  chemical bonds, 17
  chemiosmosis, 144–146
  defined, 16
  elemental structure, 15–16
  metabolism, 107–108
  photosynthesis, 122–125
  transport chains, 118–125, 146
electronegative chain, 121
element, chemical, 14–17. *See also specific elements*
embryo
  angiosperms, 279
  defined, 84
  dicot versus monocot seeds, 83–86
  germination, 84–85, 178
  liverwort life cycle, 248
  lycophytes, 258, 259
  monocot germination, 86
  moss life cycle, 252
  pines, 273
embryophyte, 239–240
enation, 261
endocarp, 93, 95
endodermis, 51, 71
endogenous rhythm, 176
endomembrane system, 35–36
endomycorrhizae, 74
endoplasmic reticulum (ER), 35
endosperm, 83–86, 279
energy. *See also* metabolism
  ATP/ADP cycle, 106–107
  biofuels, 330–332
  carbon-hydrogen bonds, 27

Index

cellular respiration overview, 10, 39, 132–134
cellular respiration processes, 134–146
chemiosmosis, 143–146
chloroplasts, 37
diffusion, 149
electromagnetic radiation, 113–114
fats and oils, 27–28
flow in ecosystem, 293–297
function of, 9
glycolysis, 134–138
importance of, 101
investment phase, 135
Krebs cycle, 139–143
metabolism process, 102
mitochondria, 39–40
payoff phase, 135
photophosphorylation, 122–125
pollution, 27
principles of, 294–295
pyramid, 295, 296
sources of, 102, 113, 118
transfer versus transformation, 117, 295, 297
transportation within plant, 10, 105–107
enzyme
 described, 22, 104–105
 function of, 22
 genetic engineering, 314–317
 Krebs cycle, 139–143
 metabolism, 104–105, 108
Ephedra (genus), 274, 334
ephedrine, 274, 334
epicotyl, 85
epidermis, 50–51, 70, 75–76
epipetalous, 284
epiphyte, 226, 227
equilibrium, 148–149
Equisetophyta (phylum), 260
ER (endoplasmic reticulum), 35
erosion, 308
estuary, 305
ethanol, 331–332
ethylene, 166, 171
etiolation, 174
etioplast, 39
eudicot, 276, 277, 282–284
eukarya, 29, 234, 239
eukaryotic cell
 basic structures, 32–35
 defined, 7

glycolysis, 135
mitosis, 185
photosynthesis, 118
eukaryotic flagella, 37
eutrophication, 310
evolution
 adaptations, 223–230
 aquatic-to-land transition, 244
 defined, 217
 hybridization, 222–223
 mutation, 218
 natural selection, 218–221
 photosynthesis, 118
 phylogeny, 234
 polyploidy, 223
 reproductive isolation, 223
 seeds, 265–266
 spores, 87
 universal ancestor, 235
exocarp, 93
extinction, 219
extracellular matrix (ECM), 41

facilitation diffusion, 150
FAD (flavin adeninine dinucleotide), 108, 143
family, chemical, 14
farming, modern, 325–326
far-red light, 174, 178
fascicle, 271
fats, 27–28
fermentation, 135, 332
fern, 196, 198, 260
ferredoxin (Fd), 124
fertilization, 9, 195
fiber, 21, 50
fibrous root system, 68–69
fiddlehead, 261
filament, 88
final electron acceptor, 121
fir tree, 268
fire, 308–309
first filial generation, 201, 202, 205
first law of thermodynamics, 294–295
flag flower, 283
flagella, 37
flavin adeninine dinucleotide (FAD), 108, 143
fleshy fruit, 93, 95
floret, 286

florigen, 176
flower
 angiosperm evolution, 277–278
 described, 79
 functions of, 8
 grasses, 286
 illustrated, 89
 legumes, 283–284
 magnoliids, 282
 orchids, 286
 single versus cluster arrangement, 90–92
 structure of, 87–89
flowering plant. *See also* angiosperm
 asexual reproduction, 184
 insect attraction, 90
 life cycles, 197
 reproduction, 87–93, 184
 seasonal responses, 175–176
 seed production, 83
 sex of, 87
 types of, 60
 vascular tissue organization, 60
fluid mosaic model, 31
fly agaric, 338, 339
fly (insect), 90, 288
fog, acid, 309
follicle, 96
food chain, 293–294
food web, 294
foot, 252
forest
 biomes, 306, 326, 327, 328, 329
 development, 254
fossil fuel, 299, 300, 309, 330
foxglove, 333
fragmentation, 184, 252
freezing temperature, 158
freshwater biome, 305
fructose, 21
fructose-1,6-bisphosphate, 138
fructose-6-phosphate, 138
fruit
 angiosperm life cycle, 280
 function of, 9
 hormones, 171
 illustrated, 93
 reproductive functions, 92–93
 seed dispersal, 97–98
 seed location, 92–93
 seed production, 83
 types of, 93–97

fruitlet, 95
fucoxanthin, 115
fumarate, 142
fungi. *See also specific types*
    blights, 303
    epidermis functions, 51
    hallucinogens, 339
    root systems, 74, 159
    symbiosis, 303

• G •

galloping moss, 344
gametangia, 247–248
gamete, 204–212, 223, 248, 264
gametophore, 248
gametophyte
    angiosperms, 279–282
    cone reproduction, 87
    conifers, 270
    evolution, 87
    gymnosperms, 267
    hornworts, 249
    lycophytes, 257–259
    mosses, 250, 252
    pines, 271–273
    plant life cycle, 195–198
    pterophytes, 262–264
    seedless vascular plants, 254, 255, 259
    seeds versus spores, 82
gamma ray, 114
Gap 1/Gap 2 phase, 186, 187
gas bubble, 158
gas exchange, 51–52, 75–76
gemmae, 184, 248
gene
    alphabetic codes, 204
    defined, 165, 199, 204
    dihybrid cross, 207–212
    flowering signals, 176
    genetic engineering, 314–317
    incomplete dominance, 212–214
    multiple-gene inheritance, 207–212
    natural selection, 220
    polyploidy, 223
    regulation, 165
    shared characteristics, 231–232
    single-gene inheritance, 199–207
generative cell, 279
genetic engineering, 311–322

genetics. *See* plant genetics
genotype, 204–212
genotypic ratio, 206, 214
genus, 240–242, 260
geotropism, 172
germination
    angiosperm life cycle, 280
    cause of, 177
    defined, 82, 177
    dicot seed, 84–85
    monocot seed, 86
    orchids, 286
    pines, 272
    process of, 178
    pterophytes, 263–264
Gibberella fujikuroi, 171
gibberellin, 166, 170–171, 178
ginkgo, 267, 268–270
global temperature, 300, 330
glucose. *See also* sugar
    cellular respiration processes, 133–138
    described, 20–21
    Krebs cycle, 139
    photosynthesis, 118
glucose-6-phosphate, 136–137
glume, 286
glyceraldedehyde-3-phosphate (G3P), 126, 138
glycolysis, 133–138
glyoxisome, 36
glyphosate, 319
gnetophyte, 267, 268, 274
Gnetum (genus), 274
golden rice, 318
Golgi apparatus, 29, 35
grain, 85, 94, 96
grana, 38
grass
    angiosperm evolution, 277
    cultivation, 222
    described, 96
    flowers, 286
    germination process, 178
    pollination, 90
grassland biome, 306–307
gravitropism, 172
gravity, 155, 172
green algae, 239–240, 244
green color, 114
green construction, 327
green revolution, 326
greenhouse gas, 27
grooming step, 140
ground meristem, 71
ground tissue system, 43, 47–50, 64

group, chemical, 14
growth. *See* plant growth/development
G3P (glyceraldedehyde-3-phosphate), 126, 138
guanine, 24, 26
guanosine triphosphate (GTP), 142
guard cell, 75, 157
gymnosperm. *See also specific types*
    versus angiosperm, 82–83, 266–267
    described, 265–266
    examples of, 267–268
    growth of, 63
    reproduction strategies, 266
    seed structure, 83
    tracheids, 54
gynoecium, 89

• H •

habitat, 292, 310
hallucinogenic plant, 337–340
haploid gamete, 191, 196, 203
haploid number, 190, 272
happy tree, 333, 334
head, 92, 284
heartwood, 65
heat, 144, 295, 297
hemlock, 335, 336
Hepatophyta (phylum), 246–249
herbaceous stem, 59
herbivore, 293
herbivory, 304
hesperidium, 95
heterosporous plant, 254, 267
heterotroph, 102–103
heterozygous organism, 204–205, 209, 212–214
hilum, 83
home building, 327
homiohydric organism, 254
homologous chromosome, 190, 192, 212
homosporous plant, 254
homozygous organism, 204–205, 212
hormone, 10, 164–171, 313
hornwort, 245, 249
horsetail, 245, 260–264
hourglass model of photoperiodism, 176
hummingbird, 90

hybrid, 205, 222–223
hydrogen
  acids and bases, 18–19
  chemiosmosis, 121, 145
  lipids, 26
  oxidation and reduction, 108
  photosynthesis, 118, 123
  transpiration, 157
hydrogen bond, 17, 25
hydrophilic/hydrophobic
    molecule, 26, 30, 148
hypertonic solution, 151
hypocotyl arch, 85
hypocotyl-radicle axis, 84, 85
hypotonic solution, 151

• I •

icons, explained, 4
incomplete dominance,
    212–214
increment borer, 65
indehiscent fruit, 96–97
indeterminate growth, 45
Indian pipe, 346
induced mutation, 218
indusium, 263
inferior ovary, 89
inflorescence, 90–91, 95, 286
inheritance
  incomplete dominance,
      212–214
  multiple genes, 207–212
  single gene, 199–207
inner membrane, 38, 40
insect
  epidermis function, 51
  pollination strategies,
      287–288, 346–347
integument, 267, 272
intercalary meristem, 46–47
inter-/intraspecific
    competition, 304
intermediate, 103–104, 135,
    138–143
intermediate filament, 37
intermembrane space, 38, 40
internode, 59, 170
interphase, 185–187, 191
ion, 15–17, 159–160
iron, 16
isocitrate, 142
isotomous branching, 256
isotonic solution, 151
isotope, 16

• J •

jimson weed, 337, 338, 339
joint fir, 274
juniper, 270
Jurassic period, 268–269

• K •

keel, 283
keratin, 37
kinetic energy, 149
kingdom, 239–240
Krebs cycle, 134, 139–143
Kurosawa, Ewiti (scientist),
    171
kwashi, 340

• L •

lab practical, 350
lab studies, 350–351
labellum, 286
lamella, 53
lamin, 37
larvae, 318–319
lateral cambia, 64
lateral meristem, 46, 59
lateral root, 72
Laws of Inheritance, 202–204,
    207
leaf abscission, 171
leaf primordia, 58–59, 75
leaf scar, 59
leaflet, 78
leaves. *See also specific types*
  aquatic adaptations, 229
  arrangements, 77, 78–79
  asexual reproduction, 184
  carnivorous plants, 228
  conifers, 270
  described, 75
  desert adaptations, 224
  dicot germination, 85
  ferns, 261
  flower arrangement, 90–91
  functions of, 8, 58
  ginkgoes, 269
  growth of, 59
  illustrated, 76
  leaf arrangements, 78–79
  lycophytes, 256
  parts of, 75
  phylogenetic tree, 236

  pine tree, 271
  seedless vascular
      plants, 254
  structure of, 75–76
  tissue organization, 75–76
  transpiration, 156
  tropical adaptations,
      226, 228
  turgor movements, 173
  types of, 77–80
  vegetables, 78
lecture, 349–350
legume, 94, 96, 283–284
lemma, 286
lenticel, 52, 64
lettuce, 170, 178
leucoplast, 39
lichen, 303
light
  adaptations, 225, 226
  germination, 178
  phototropism, 172
  seasonal responses,
      174–178
  spectrum, 113–115
light independent reaction,
    116–118, 125–129
light reaction, 116–124
lignin, 41
lily, 285
linen, 50
linking step, 140
Linnaeus, Carl (scientist), 240
lipid, 27–28
listening skills, 349–350
liverwort, 196–197, 245–250
locoweed, 336
locus, 204
lodicule, 286
long day plant, 175
lycophyte, 255–259
lycopod, 245, 255
lysosome, 29, 36

• M •

macromolecule, 19–28
magic mushroom, 338, 339
magnolia tree, 276, 277, 278
magnoliid, 277, 282–283
malate, 139, 142
mandrake, 339–340
maple tree, 276, 277
marijuana, 338, 339
marine biome, 305

mass number, 16
matrix, 40
matter
  components of, 13–18
  cycles of, 298–301
  defined, 9, 13, 102
  metabolic pathways, 103
mechanical dispersal, 97, 98
medicine, 271, 320, 333–334, 339
megagametophyte, 279
megapascal, 152–153
megaphyll, 254
megasporangia, 257, 259
megaspore, 254, 259, 272
meiocyte, 272
meiosis
  chromosome assortment, 211–212
  chromosome number, 190–191
  defined, 11
  evolution, 223
  gene assortment, 211
  liverwort, 249
  lycophytes, 259
  versus mitosis, 191
  mosses, 252
  phases of, 191–194
  pines, 272
  plant life cycle, 195–198
  pterophytes, 262–263
  spores, 82
meiospore, 196
melatonin, 177
memory, 352
Mendel, Gregor (scientist), 200–204
meristem, 43–47, 249
meristematic tissue, 44, 71
meristematic tissue system, 43
mesocarp, 93, 95
mesophyll, 76, 128–129
metabolic pathway, 103–104
metabolism. *See also* energy
  ATP/ADP cycle, 105–107
  cellular respiration, 133
  described, 101–103
  electron transfer, 107–108
  enzymes, 104–105
  illustrated, 102
  metabolic pathways, 103–104
  types of, 101–102
metaphase, 193, 194, 198
methane, 332
microfilament, 37

microgametophyte, 279
microphyll, 254
micropyle, 83, 272
microsporangia, 257
microspore, 254, 259, 267
microsporophyll, 257
microtubule, 37
middle lamella, 41
midrib, 250
milkweed, 96, 319
Mimosa pudica, 173
mineral nutrient, 159–160
mitochondria
  cellular respiration, 134, 140
  chemiosmosis, 145
  described, 39–40
  electron transport chain, 146
  illustrated, 29, 40
mitosis
  angiosperms, 279–280
  cytokinesis, 189
  described, 11, 182, 185, 187
  interphase, 185–187
  liverwort, 247
  lycophytes, 258
  mosses, 251–252
  phases of, 185–189
  plant life cycle, 195–198
  pterophytes, 264
  root systems, 71
mitotic spindle, 187, 191
mnemonic, 238, 352
mold, 81
molecule
  acids and bases, 18–19
  active transport, 150–151
  cell construction, 19–28
  cellular respiration, 132
  diffusion, 149–150
  electron carriers, 107–108
  formation of, 17–18
  heat response, 144
  metabolic pathways, 103–104
  transfer across plasma membrane, 148–149
monocot
  angiosperm life cycle, 280
  described, 60, 285–286
  flower structure, 88–89
  germination, 86
  leaf tissues, 76
  roots, 70
  seed structure, 83, 85–86
  stems, 60–61
monoecious plant, 248

monohybrid, 202–207, 214
monosaccharide, 20–21
Mormon tea, 274, 334
moss, 245, 250–253
moth, 90, 288, 318
mountain forest, 306
mouse-ear cress, 206
movement, plant, 172–173, 177
mucilage, 249
multiple fruit, 95
mushroom, 338, 339
mutation, 185, 218, 204
mutualistic symbiosis, 74, 303, 304
mycorrhizae, 74, 159, 346

NAD+. *See* nicotinamide adenine dinucleotide
NADP+. *See* nicotinamide adenine dinucleotide phosphate
nanometer, 114
natural selection, 218–221
near-basal meristem, 249
nectary, 302
negative charge, 15
negative geotropism, 172
negative pressure, 158
negative thigmotropism, 173
neriine, 336
netted vein, 76
neutron, 15–16
nicotinamide adenine dinucleotide (NAD+)
  chemiosmosis, 143, 145
  glycolysis, 135, 138
  Krebs cycle, 140, 142
  metabolism, 108
nicotinamide adenine dinucleotide phosphate (NADP+)
  cellular respiration, 133, 135, 138
  Krebs cycle, 140
  metabolism, 108
  photosynthesis, 117, 118, 121–129
nightshade, 284, 335, 336
nitrate, 300, 301, 310
nitrification, 301
nitrogen cycle, 300–301
nitrogen fixation, 269, 300, 303
nitrogenous base, 24–25
node, 46, 59, 236

noncyclic
  photophosphorylation, 122–124
nondisjunction, 223
note taking, 349–350
nuclear membrane, 32, 187, 189
nucleic acid, 23–26
nucleoid, 32
nucleoli, 187
nucleotide, 23–24
nucleus, 29
nut, 94, 96

• O •

oak tree, 65
oil, 332
oleander, 335, 336
oleoplast, 39
oligosaccharide, 20–21
operculum, 252
Ophioglossum, 190
opposite leaf, 77, 79
orchid
  adaptations, 226
  described, 286
  pollination, 288, 346–347
  roots, 73
organ. *See specific organs*
organelle, 33
osmosis, 151–152, 162
osmotic potential, 153
outcrossing plants, 201
outer membrane, 38, 40
outgroup, 236
ovary
  angiosperms, 280
  defined, 9, 88
  flower structure, 88, 89
  fruit structure, 92, 93
  fruit types, 95
ovule
  angiosperms, 278, 279–282
  defined, 9, 82, 88
  fruit structure, 83, 92
  gymnosperms, 267
  pines, 272, 273
oxaloacetate, 139, 142–143
oxaloacetic acid, 129
oxidation
  described, 107–108
  glycolysis, 133
  Krebs cycle, 139–143
  photosynthesis, 116
  pyruvate, 140

oxidative phosphorylation, 134, 143, 145
oxygen
  cellular respiration overview, 133
  elemental structure, 15
  epidermis functions, 51
  Krebs cycle, 139–143
  nitrogen cycle, 301
  photosynthesis, 116, 123–125
ozone layer, 309

• P •

palea, 286
palisade parenchyma, 76
palm, 269
palmately compound leaf, 77–78
panicle, 91–92
paper making, 328–330
parallel vein, 76
paraphyses, 252
parasitism, 304
parenchyma
  described, 47–48
  dicot stems, 61–62
  illustrated, 48
  leaf structure, 76
  liverwort, 247
  periderm, 52
  phloem, 55
  vessels, 53
parental generation, 201–202, 207–214
passive transport, 150
PC (plastocyanin), 123, 124
peach, 93
peanut, 83–84
peas, 170, 190, 200–207
peat moss, 253
pectin, 41
pedicel, 90
peduncle, 87
PEP carboxylase, 127, 129
pepo, 95
peptide bond, 22
perennial plant, 63–65
perianth, 88
pericarp, 85, 95, 96
pericycle, 72
periderm, 51–52, 64
Periodic Table of Elements, 14
peristome, 252

peroxisome, 36
pesticide, 319
petal, 88, 283, 286
petiole, 75
peyote, 338
PGA (3-phosphoglycerate), 126
pH, 18–19
pharma crop, 320
phellogen, 46, 64, 72
phenotype, 204–214
phenotypic ratio, 206, 214
pheophytin, 123
phloem
  described, 10, 54–55
  energy transport, 10
  fiber, 55
  parenchyma cell, 55
  ray cell, 55
  secondary growth, 64
  sugar-transport mechanisms, 160–162
  vascular bundles, 60
phosphate group, 24–26, 105–106, 122
phosphoenolpyruvate, 138
phospholipid, 28, 30, 38
phosphorylation, 122, 143
photolysis, 123
photoperiod, 175, 177
photophosphorylation, 122–125
photoreceptor, 172, 174
photorespiration, 127
photosynthesis
  absorption spectra, 114–115
  basic process, 9, 37
  carbon dioxide reduction, 116
  carnivorous plants, 228
  versus cellular respiration, 131–132
  chemical reaction, 112, 116
  described, 37, 102, 112
  evolution of plants, 118
  function of, 113
  illustrated, 117
  light independent reaction, 116–118, 125–129
  light reaction, 116–124
  pigments, 114–115
  plant structure, 8
  role of soil, 109–112
  solar energy, 113–114
  steps of, 109
  water oxidation, 116

photosystem, 119
Photosystem I/II (light reaction component), 122–125
phototropism, 172
photovoltaic cell, 119
phycobilin, 115
phycocyanin, 115
phycoerythrin, 115
phylogenetic tree, 233–236, 277
phylogeny, 234, 240
phylum, 238
phytochrome, 174–176, 178
pigment
  absorption spectra, 114
  defined, 114
  energy transfer, 119
  light reactions, 122–125
  types of, 115
pine tree, 66, 268, 271–273
pineapple, 95
pinnately compound leaf, 77–78
pistil, 88, 95
pit, 52–53, 95
pitcher plant, 228, 229
pith, 61
plant. *See also specific plants*
  basic needs of, 3
  diversity of, 11
  fossils, 54, 244, 277, 299
  functions of, 7, 9–10, 12
  movements, 172–173, 177
  structure of, 7–9
plant cell. *See* cell, plant
plant genetics
  agamospermy, 183
  asexual reproduction, 185
  coding, 204
  defined, 199
  dihybrid cross, 207–212
  versus human genetics, 195
  incomplete dominance, 212–214
  multiple-gene inheritance, 207–212
  mutations, 218
  natural selection, 218–221
  overview of, 11
  sexual reproduction, 184, 185
  single-gene inheritance, 199–207
  terminology, 204–205
plant growth/development
  differentiation, 163–164
  growth movements, 172–173
  hormones, 165–171

overview of, 163–165
seasonal responses, 174–178
seedless vascular plants, 254
plant life cycle
  angiosperms, 279–281
  described, 194–198
  hornworts, 249
  liverwort, 196–197, 247–249
  lycophytes, 257–259
  mosses, 251–252
  pines, 272–273
  pterophytes, 262–264
plant reproduction. *See also specific reproduction processes*
  adaptations, 244
  asexual, 181–184
  bryophytes, 246
  cones, 87
  conifers, 270
  cycads, 269
  flowering plants, 87–93
  fruit, 92–93
  gymnosperms, 266
  hybridization, 222
  methods of, 81
plant reproduction *(continued)*
  natural selection, 220–221
  overview of, 11
  plants versus animals, 194–195
  pterophytes, 262–263
  seeds, 82–86
  sexual, 184–185
  spores, 81–82
plantlet, 184
plasma membrane, 28–32, 71, 148–149
plasmodesmata, 42, 129
plasmolysis, 154
plastid, 38–39
plastocyanin (PC), 123, 124
plastoquinone (PQ), 123, 124, 125
ploidy, 190
plumule, 84, 85
pneumatophore, 73
poikilohydric organism, 246
Pointsettia plant, 79
poisonous plant, 284, 334–337
polar covalent bond, 17, 18
polar nuclei, 279, 280
pollen
  angiosperms, 280, 283, 285
  flower structure, 88

grain, 267
source of, 9
tube, 267, 280
pollination
  angiosperm strategies, 278, 286–288
  cycads, 269
  defined, 9
  flower structure, 88
  genetically engineered crops, 321
  insect-attracting flowers, 90, 144, 346–347
  pines, 272
pollinator, 144. *See also specific pollinators*
pollinia, 286
pollution, 27, 309–310, 329
polynucleotide chain, 24–26
polypeptide chain, 22
polyploid, 190
polyploidy, 223
polysaccharide, 20–21
pome, 94–95
population, 291–292
pore, 247
positive charge, 15
positive geotropism, 172
positive phototropism, 172
positive thigmotropism, 173
potato, 183
PQ (plastoquinone), 123, 124, 125
predator, 297, 304
prescribed burn, 309
pressure potential, 153
pressure-flow hypothesis, 160–162
primary cell wall, 41
primary consumer, 293, 296
primary endosperm, 83
primary growth, 59–62
primary productivity, 308
primary root, 85
primary succession, 307
procambium, 71
producer, 293, 296, 298, 300
prokaryotic cell, 32, 135
prometaphase, 187
prop root, 8, 73, 86, 226–227
prophase, 187, 192–193, 194
proplastid, 39
protein. *See also specific types*
  active transport, 151, 152
  chemiosmosis, 145
  circadian rhythm, 177

creation, 32, 35
cytoskeleton, 36–37
described, 22–23
facilitation diffusion, 150
flowering signals, 176
hormonal responses, 164–165
plasma membranes, 31
ribosomes, 32, 35
proteinoplast, 39
protoderm, 71
protogynous reproduction, 263
proton, 15–16, 123, 125
protonema, 247, 251–252
protoplast, 31
provitamin A, 318
pseudobulb, 286
Psilophyta (phylum), 260
Pterophyta (phylum), 260–264
pulp, 329
pulvinus, 173
pump protein, 151
Punnett square, 205–211, 213
pure-breeding, 201
pyruvate, 133–143

• Q •

Queen Victoria water lily, 347
question, test, 353
quillwort, 245, 255

• R •

raceme, 91–92
rachilla, 286
rachis, 90
radial symmetry, 250, 282
radicle, 68, 84, 85
rain, 159, 309
rainforest biome, 306, 326
rainforest plant, 226–227, 302
ray cell, 53, 65
ray flower, 92, 284
reaction center, 119–120
receptacle, 92, 95
receptor protein, 23, 31, 164
recessive gene, 203, 212
recombinant DNA technology, 311
recycling, 330
red color, 288
red light, 174, 178
redox reaction, 107
reduced electron, 107, 108
reduction
  cellular respiration, 133
  described, 107–108
  Krebs cycle, 139, 140
  photosynthesis, 116
redwood, 268
regeneration, 126
region of differentiation/elongation, 71
replicated chromosome, 186
reproduction. See plant reproduction
reproductive adaptation, 244
reproductive isolation, 223
reproductive leaf, 79–80
RER (rough endoplasmic reticulum), 35
resin canal, 271
restriction enzyme, 314–316
resurrection plant, 345
reticulate vein, 76
rhizoid
  liverwort, 247
  mosses, 250, 252
  pterophytes, 263–264
rhizome, 66–67, 256, 263
rhizophore, 72, 256
ribonucleic acid (RNA), 23–26, 32, 235
ribose, 24
ribosomal RNA (rRNA), 32, 232
ribosome, 32, 35, 232
ribulose bisphosphate (RuBP), 125–126
rice, 171, 318
rind, 95
ripening fruit, 171
root
  asexual reproduction, 183
  cap, 70–71
  cavitation, 159
  components of, 71
  cycads, 269
  functions of, 8, 58, 68
  geotropism, 172
  growth of, 69, 164, 170
  hair, 51, 70, 72, 159
  pressure, 159
  soil contact, 159
  sugar storage, 160
  thigmotropism, 173
  transpiration, 156
  transport control, 71
  types of, 159
root system
  adaptations, 226–227
  defined, 44, 68
  dicots, 70–72
  ferns, 261
  fungi, 74, 159
  illustrated, 45
  lycophytes, 256
  mitosis, 71
  seedless vascular plants, 253
  types of, 68–69
rough endoplasmic reticulum (RER), 35
Roundup (weedkiller), 319
rRNA (ribosomal RNA), 32, 232
rubisco, 125–129
RuBP (ribulose bisphosphate), 125–126
runner, 66

• S •

S (synthesis) phase, 186
salicin, 333
saltwater, 305
samara, 94, 97
sapwood, 65
saturated fat, 27–28
scale, 247
scaly leaf, 80
scavenger, 293
scent, 144, 287, 288, 344
schizocarp, 94, 97
scientific name, 241
sclereid, 50
sclerenchyma, 48–50
scopolamine, 336
seasonal response, 174–178
seaweed, 115, 122
second filial generation, 202
second law of thermodynamics, 295
second messenger, 164
secondary cell wall, 41
secondary consumer, 293, 296
secondary growth, 46, 63–65, 72
secondary phloem, 64–65
secondary succession, 307
secondary xylem, 64
secretory cell, 47
secretory vesicle, 36
seed
  coat, 83–85, 173–178, 273
  conifers, 270
  cycads, 269

root *(continued)*
  defined, 8
  dispersal, 97–98
  dormancy, 177
  dry fruits, 96
  evolution, 265–266
  fruit structure, 92–93
  germination process, 177–178
  germination time, 82
  ginkgoes, 269, 270
  gymnosperms versus angiosperms, 82–83
  orchids, 286
  pines, 273
  versus spores, 82
  structure of, 83–86
seedless vascular plant, 244–245, 253–255. *See also specific types*
selection pressure, 221
self-crossing plants, 201
self-fertile plant, 201, 264
sepal, 87–88
SER (smooth endoplasmic reticulum), 35
serotonin, 339
sessile flower, 92
seta, 250–251
sexual reproduction, 184–185, 247–248
shoot system, 44–45, 172–173, 261
short day plant, 175
sieve plate, 54–55
sieve-tube elements, 54–55, 160–162
signal transduction, 31, 164, 173
silicle, 96
silique, 94, 96
simple diffusion, 150
simple fleshy fruit, 95
simple leaf, 77–78
simple ovary, 88
simple tissue, 47
sink, 160, 162, 326
sister chromatid, 186, 191, 192
sister group, 236
skotoperiod, 175, 176
smooth endoplasmic reticulum (SER), 35
sodium, 15, 16
soft wood tree, 271
soil, 109–112, 159–160
solar energy, 113–114, 118, 119
solar panel, 119
solar tracking, 75

solute, 151–153, 160
solution, 18–19, 151
somatic cell, 203
sori, 263
source, sugar, 160
soybean, 175, 319
speciation, 219
species
  binomial nomenclature, 240–242
  defined, 219
  ecological role, 292–293
  relationships between, 232
specific epithet, 240–242
sperm
  angiosperms, 279, 280
  cycads, 269
  described, 9
  ginkgoes, 270
  liverwort life cycle, 248
  lycophytes, 258, 259
  moss life cycle, 252
  pines, 272
  plants versus animals, 195
  pterophytes, 264
  role of, 184
  trait predictions, 205–207, 209–211
Sphagnum moss, 253
spike leaf, 92
spike moss, 255, 257–259
spikelet, 286
spindle fiber, 187
spine, 79–80
spongy parenchyma, 76
sporangia
  defined, 87
  flower structure, 87–88
  gymnosperms versus angiosperms, 82
  moss, 252
  pines, 272
  pterophytes, 262
  seedless vascular plants, 255, 257–259, 262
spore
  described, 8, 81–82
  evolution of, 87
  growth of, 191
  liverwort life cycle, 249
  lycophyte life cycle, 257–259
  meiosis, 82, 190
  moss life cycle, 252
  mother cells, 252
  plant life cycle, 196

pterophytes, 263–264
seedless vascular plants, 254
versus seeds, 82
sporophyll, 87, 257
sporophyte
  angiosperms, 279–282
  liverwort, 196, 248–250
  lycophytes, 256
  pterophytes, 262–264
  seedless vascular plants, 254
spring wood, 65
spurge flower, 283
stamen, 88, 284, 286
standard, 283
star anise, 277
starch, 20, 21, 126
starch-statolith hypothesis, 172
stem
  asexual reproduction, 183
  described, 8, 58–59
  flower arrangements, 90–91
  growth of, 59–65
  specialized, 66–67
stem cell, 44
steroid, 28
sticky end, 315
stigma, 88, 286
stilt root, 226
stipule, 75
stolon, 66–67
stoma, 75
stomata
  defined, 51
  hornworts, 249
  light independent reactions, 126–128
  transpiration, 157
  water loss, 170
stone cell, 49
stone plant, 225, 344–345
storage root, 73
strawberry, 95, 183
stress, 12
strobili, 87, 257, 262, 272
stroma, 38
study time, 351–352
style, 88
suberin, 52, 64, 71
substrate, 103–105, 135–136
substrate-level phosphorylation, 135–138, 142
succession, 307–310
succinate, 142
succinyl-CoA, 142

Index

succulent
  adaptations, 224–225
  described, 79
  illustrated, 80
  photosynthesis, 127–128
sucrose, 20
sugar. *See also* glucose
  cellular respiration overview, 132–133
  described, 20, 21
  light independent reactions, 125–129
  nucleotides, 24
  photosynthesis, 117, 118, 125–129
  polynucleotide chains, 26
  storage, 160
  transportation within plant, 10, 160–162
summer wood, 65
sun
  energy production, 118
  importance of, 294
  phototropism, 172
  seasonal responses, 174–178
  solar energy, 113–114, 118, 119
  solar tracking, 75
sundew, 228, 229
sunflower, 96, 276, 277, 284
superior ovary, 89
superweed, 320
supporting material, 353
surface tension, 155
Sustainable Cotton Project, 330
sustainable product, 327
symbiosis, 302–304
synthesis (S) phase, 186
systematics, 236, 237

• T •

taiga, 306
tank plant, 226, 302
tap root system, 68–69, 159
taxol, 271, 334
taxon, 234, 236
taxonomic hierarchy, 236–242
Taxus brevifolia, 271, 334
teaching assistant, 351, 354
telophase, 189, 194
temperate deciduous forest, 306
temperature, environmental, 157–158, 176–177, 224, 300

10 percent rule, 297
tendril, 67, 79–80, 173
tension, 158
tepal, 88, 286
terminal bud, 58–59, 87–88
tertiary consumer, 293, 297
test taking, 353, 354
tetrad, 192
thallus, 247, 249
Theophrastus (scientist), 240
thigmotropism, 173
3-phosphoglycerate (PGA), 126
thylakoid, 38, 123, 125
thymine, 24, 26
Ti plasmid, 314–317
tissue
  cultures, 169–170, 313
  defined, 43
  formation, 43
  systems, 43–55
tonoplast, 36
top predator, 297
touch response, 173
tracheid, 52–54, 158
transcription, 32
transcription factor, 176
transgenic *Bt* corn, 319
transgenic organism, 311
translation, 32
transpiration, 156–158
transport protein, 23, 31, 149, 152
transport tissue, 52–55
transport vesicle, 36
transporting material
  active transport, 150–151
  cavitation, 158–159
  cohesion of water, 154–156
  described, 10
  diffusion, 149–150
  illustrated, 148
  importance of, 147
  osmosis, 151–152
  plasmolysis, 154
  pressure-flow hypothesis, 160–162
  proteins, 23
  roots, 159–160
  selective permeability, 148–149
  sugar transfer mechanisms, 160–162
  transpiration, 156–158
  turgor pressure, 152–153

tree, 63–66, 329. *See also specific types*
trichome, 50–51
triglyceride, 27–28
triploid, 83, 279
Triticum monococcum, 222
trophic level, 293
trophic pyramid, 295
tropical plant, 226–227
tropism, 172–173
true berry, 95
true dicot, 282–284
true fern, 245, 260–264
tube cell, 279
tuber, 67
tubulin, 37
tulip tree, 276, 277
tundra biome, 307
turgor pressure, 152–153, 160

• U •

umbel, 91–92
umbellate, 92
universal ancestor, 235
unsaturated fat, 27–28
uracil, 24

• V •

vacuole, 29, 36
vascular bundle, 60–61, 64, 70
vascular cambium, 46, 64, 72
vascular cylinder, 70
vascular tissue system
  defined, 44
  leaf components, 76
vascular tissue system *(continued)*
  secondary growth, 64, 72
  seedless vascular plants, 254
  tissues types in, 52–55
vegetative adaptation, 244
velamen, 51, 286
Venus fly trap, 228, 229
vernalization, 176
vesicle, 35–36
vessel, 53–54
vine, 226, 227, 274
visible light, 113–115
Vitamin A, 318
volcanic eruption, 307

• W •

wasp, 288, 346–347
water
  absorption, 51, 71, 74
  adaptations, 197–198, 229–230
  biomes, 305
  carbon cycle, 298–299
  cavitation, 158–159
  cellular respiration, 132, 142
  cohesion for transport, 154–156
  formation of, 18
  Krebs cycle, 142
  lipid structure, 26
  loss from cells, 50–51, 71, 154
  osmosis, 151–152
  paper making, 329
  photophosphorylation, 123–124, 131–132
  photosynthesis, 116, 120, 123–124
  potential, 152–153, 156
  seed dispersal, 97
  seedless vascular plants, 253
  storage of, 224
  stress, 170
  sugar transport, 160–162
  transpiration, 154, 156–158
  turgor movements, 173
  turgor pressure, 152–153
water hyacinth, 229, 230
water lilly, 229, 230, 277, 347
wavelength, 114, 115
wax, 38, 226, 227
weed killer, 319–320
Welwitschia plant, 274, 345–346
Went, Frits (botanist), 166–168
wet milling, 332
wetland, 12
wheat, 222, 326
whisk fern, 245, 260–264
white willow, 333
whorl, 87–89, 91
whorled leaf, 79
wildland fire, 308–309
wild-type trait, 204
wind, 90, 97–98, 157, 287
wing, 283
winged fruit, 97
Woese, Carl (scientist), 235
wood
  cutting boards, 68
  defined, 65
  home construction, 327
  paper making, 328–330
  types of, 65
woody dicot, 72, 158
woody stem, 59, 63–65
wormwood, 333

• X •

xylem
  described, 10, 52–54, 154
  energy transportation within plants, 10
  illustrated, 53
  ray cells, 55
  secondary growth, 64
  vascular bundles, 60
  water transport, 154–159

• Y •

yew, 270, 271, 334

• Z •

Z-scheme, 122–124
zygomorphic flower, 282
zygote, 195, 248, 252, 258–259